A Short Introduction to
STATA 8 FOR BIOSTATISTICS

www.**timberlake**.co.uk

A Short Introduction to
STATA 8 FOR BIOSTATISTICS

Michael Hills

Bianca L. De Stavola

Timberlake Consultants Ltd.

A Leading Distributor of Statistical and Econometrics Software

A Short Introduction to Stata 8 for Biostatistics
Copyright © Michael Hills and Bianca L. De Stavola, 2002

British Library Cataloguing-in-Publication Data
A catalogue record for this book is available from the British Library

First edition 2003, published by Timberlake Consultants Press a Division of Timberlake Consultants Ltd.

Cover designed in the UK by CCA Design & Communications Ltd, London
Printed in the UK by Allstar Services Ltd

ISBN 0-9542603-1-7

Published by:
Timberlake Consultants Ltd
Unit B3, Broomsleigh Business Park, Worsley Bridge Road, London SE26 5BN, U.K.
http://www.timberlake.co.uk
http://www.timberlake-consultancy.com

Preface

This book is a revised version of a book with a similar title which was prepared for Stata 7. The datasets and programs have been updated to Stata 8, and the book has been expanded to include new chapters on dialog boxes, the new graph commands, likelihood ratio tests, and Mantel-Haenszel methods. Several of the original chapters have also been expanded.

Starting to use Stata is relatively simple, but because of its size, and the wealth of information in the guide and manuals, the question of what comes next can be rather daunting. This book provides a short introduction which will help answer this question. Although written with biostatisticians in mind, much of the material in the book is equally relevant to other disciplines.

We believe that the only way to learn Stata is to try it out, so we assume that the reader is seated in front of a computer which is running Stata. For this reason we have not felt it necessary to print the output which follows Stata commands. Of course there are occasions when it is reassuring to see that the output on the screen is the same as the output on the page, and we occasionally include some output for this reason, but the spirit of the book is to try something and see what happens. In keeping with this we have not provided solutions to the exercises, but instead a program produces the solutions on the screen.

We have mostly used official Stata commands but, as one of the features of Stata is the large number of user–contributed commands, we have also felt free to use a few of these, where appropriate. In particular, Chapters 10 – 12 depend on two commands, written specifically for this book, which provide dialog boxes for making tables and estimating effects.

The datasets and additional program files are an integral part of the book. They are included on the CD-ROM which comes with the book, but are also available as a download from *www.timberlake.co.uk* or *www.stata.com*. Instructions on how to proceed with each of these media are given in Chapter 0: Getting Started. We would be grateful to be notified of any errors: as and when the need arises, updates and an errata file will be included on the websites.

The ideas in Chapters 10 – 12 arose from courses taught jointly by Michael Hills and David Clayton, and we gratefully acknowledge David's contribution. We are also grateful to Nick Cox for reading a draft of the Stata 7 book and making many helpful suggestions.

Michael Hills, Retired
email: *mhills@blueyonder.co.uk*

Bianca De Stavola, London School of Hygiene and Tropical Medicine
email: *bianca.destavola@lshtm.ac.uk*

December 16, 2003

Contents

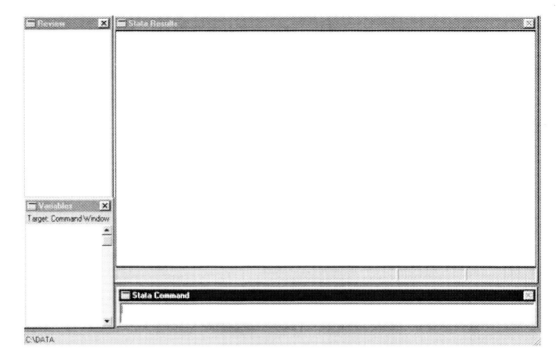

Figure 1: The four Stata windows and the name of the working directory

Chapter 0

Getting started

When you start Stata you will see the four windows shown in Figure 1:

```
Review
Variables
Stata Results
Stata Command
```

The working directory, that is the directory where Stata expects to find the data when no path is specified, is shown at the foot of the screen, on the left. In Figure 1 it is `C:\data`, which is usually the default working directory.

Typing and editing commands

Commands are typed in the command window. When typing commands, remember that Stata is case sensitive, so 'A' is not the same as 'a'. To edit a previously run command, click on it in the review window, or use the Page-Up key to retrieve the previous command line. Then modify it in the command window, as required.

Stata prompt

When a command is executed it will appear in the results window with a dot in front. The dot is there to distinguish between commands and results, and is referred to as the Stata prompt. In this book we shall indicate those commands which you need to type into the command window by starting them with a Stata prompt. You should not type the prompt - only the command. For example,

```
. describe
```

means you should type describe in the command window.

Files for the book

An integral part of the book is a set of data files and programs. They need to be installed in a convenient folder. For MS-Windows users we suggest making a folder hs8 within the C:\data folder which should already exist on your hard drive (C:\). You can choose another name for this folder if you prefer, but if you do you will need to replace hs8 by this new name in what follows.

- To load the files from the CD-ROM, start Stata and type the following commands in the Command window (don't type the Stata prompt)

```
. cd C:\data\hs8
. net from D:\
. net install book8
. net get book8
```

On some PC's the CD-ROM is referred to by another letter, such as E, instead of D.

- Watch out for error messages. If files with the same names have already been installed, and their content is different from that of the files on the CD-ROM, Stata will display an error message and will not install the new files. To overwrite the old files with the new files you need to type

```
. net install book8, replace
. net get book8, replace
```

- To download the files from the Timberlake website, start Stata and type the following commands in the Command window:

```
. cd C:\data\hs8
. net from http://www.timberlake.co.uk/data/hs8/
. net install book8
. net get book8
```

- To download from the Stata website, replace `www.timberlake.co.uk` with `www.stata.com`.

Stata under Unix or Macintosh

If you are running Stata under Unix or Macintosh you need to change these instructions to refer to the working directory and CD-ROM under the Operating System you are using.

Fonts

The default font for each of the Stata windows can be changed. To change the font for the results window when using MS-Windows, right click anywhere in the Stata results window. This will bring up a menu which allows you to increase or decrease the size of the font, and also to change the font style.

Updating your version of Stata

Stata 8 is not supplied with this book, and you may be using a version which is older than the most recent one. It is always a good idea to keep your version up-to-date, and this is easy if you have an internet connection. Simply enter

```
. update all
```

in the command window and then follow the instructions. Chapter 20 gives more details about updating Stata.

Getting out of Stata

Select the *File* tab on the top of the screen and then select *Exit*. To interrupt a Stata command, click on the *Break* icon (a red circle with a white cross on the top of the screen: see Figure 2).

How to read this book

The book is meant to be read while sitting in front of a computer running Stata 8. The material is quite dense, and one or two chapters at a sitting is probably enough

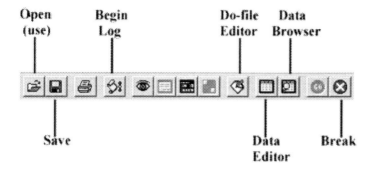

Figure 2: The Stata icons at the top of the screen

to exhaust most people's concentration. Comments about a command generally come before the command, but sometimes a few words of explanation will follow, so it is always worth checking the next line of text when something unexpected happens. Try to avoid the temptation to hop from one Stata command to the next – the words between can be important!

It is important to work through both Chapter 1 and Chapter 2, even though they cover the same material. Chapter 1 asks you to type commands in the command window, and Chapter 2 uses menus and dialog boxes. Working through both chapters will give you a good idea of the strengths and weaknesses of the two approaches.

Chapter 1

Some basic commands

This chapter introduces some of the basic Stata commands which are used regularly. Topics covered include how to load a dataset which is already in Stata format (Chapter 4 deals with other types of format); how to find out the names of the variables in the dataset; how to list and browse the data values for one or more variables; how to make simple tables; how to generate new variables; how to sort; and how to use Stata as a calculator.

1.1   The births data

Variable	Units or Coding	Type	Name
Identity number	–	categorical	id
Birth weight	grams	metric	bweight
Birth weight < 2500 g	1=yes, 0=no	categorical	lowbw
Gestational period	weeks	metric	gestwks
Gestational period < 37 weeks	1=yes, 0=no	categorical	preterm
Maternal age	years	metric	matage
Maternal hypertension	1=hypertensive, 0=normal	categorical	hyp
Sex of baby (numeric)	1=male, 2=female	categorical	sex
Sex of baby (alphabetic)	male, female	categorical	sexalph

Table 1.1: Variables in the births dataset

To introduce Stata we use the births dataset which concerns 500 mothers who had singleton births in a large London hospital. Some of the variables in this file are categorical, taking different categories as values; others are metric taking measurements as values. In general it is better to code categories numerically, but in order to demonstrate some features of Stata the sex of the baby is coded using numbers in the variable `sex` and using alphabetical characters in the variable `sexalph`. A sequence of characters such as "male" or "year 1990" is called a *string* and variables that hold such characters are called *string variables*. Variables that hold numerical data are called *numeric variables*. In the births dataset `sex` is a numeric variable while `sexalph` is a string variable.

1.2   A first look at the data

In the births file each record contains the variable names and values for one of the 500 mothers. In Stata terminology these records are also called observations. To load the data, first check that your working directory is correct with the command

. pwd

which stands for print working directory (don't type the Stata prompt). For MS-Windows this should show

C:\data\hs8

If you are not in the correct working directory, try

. cd C:\data\hs8

to change the working directory. Now type

. use births

The results window will show this command followed by its results, in this case the label of the dataset. The review window will also show the command for future reference. If you get the error message 'Dataset not found' it means one of two things: either you have not loaded the files from the CD correctly, or you have not set the working directory correctly. You can check which files are actually in the working directory with the command

. dir

Once you have successfully loaded the births dataset you can browse its content in a spreadsheet by clicking on the *Data Browser* icon (a sheet of printing with a magnifying glass, at the top of the screen) or typing

. browse

in the command window. Close the browse window when you have finished browsing. You can see the names of the variables in the variables window, but a more informative way of doing this is by typing

```
. describe
```

in the command window.

A good way to start an analysis is to ask for a summary of the data by typing

```
. summarize
```

This will produce the mean, standard deviation, and range, for each variable in turn, provided it is not a string variable. For string variables the summary is left blank. For a more detailed summary of the variable `gestwks` try

```
. summarize gestwks, detail
```

which lists the four smallest values and the four largest values, together with various summaries of the distribution. Listing the smallest and largest observations is a useful check on data errors.

In most datasets there will be some missing values. These are usually coded using the symbol `.` in place of the value which is missing if the variable is numeric, or as blank if the variable is string. If you wish to make a distinction between types of missingness, you can use up to 26 other missing values codes. These are `.a`, `.b`, `.c`, up to `.z`, but it is unlikely that you will need them all!

Browse the dataset again and you will notice that there are two missing values in line 2. The `codebook` command is also useful for seeing whether there are missing values:

```
. codebook gestwks
```

The output shows, on the right-hand side, that there are 10 missing values for this variable.

The `list` command is used to list the values in the data file. Try out the following and see the consequences:

```
. list matage
```

Stata stops after each screen of output. Click on the *Go* button (a green circle at the top of the screen with the word GO on it), or press the space-bar, to get another screen. Alternatively, press the ⏎ key to continue line by line. You can cancel this command (and any other Stata command) by clicking on *Break* (the red ⊗ icon at the top of the screen).

Stata commands can be restricted to observations 1, 2, ..., 5 (for example) by adding `in 1/5` to the command. Try

```
. list matage in 1/5
. list matage bweight in 11/20
```

The command `list` with no variable names, will list the data for all the variables in the dataset. For example,

```
. list in 1/5
```

lists all variables for observations 1–5. Depending on the width of your screen the data are listed either in a display or table style. You can however force one or the other by adding the relevant option to the command. Try them both to see the difference:

```
. list in 1/5, display
. list in 1/5, table
```

If you wish to remove all dividing lines, try

```
. list in 1/5, clean
```

The observation number can be omitted by using the option noobs, as in

```
. list in 1/5, table noobs
```

1.3   Tables of frequencies

When starting to look at any new dataset the first step is to check that the values of the variables make sense and correspond to the codes defined in the coding schedule. For categorical variables this can be done by looking at one-way frequency tables and checking that only the specified codes occur. The frequencies of the values taken by the categorical variables hyp and sex can be viewed by typing

```
. tabulate hyp
. tabulate sex
```

Their cross-tabulation is obtained by typing

```
. tabulate hyp sex
```

Cross-tabulations are useful when checking for consistency. The basic output from a cross-tabulation reports frequencies only; to include relative frequencies (percentages) for rows and/or columns add the options row and col as in

```
. tabulate hyp sex, row
. tabulate hyp sex, col
```

If you wish to see only the percentages use the option nofreq. To examine the frequency of missing values use the option missing with the tabulate command.

```
. tabulate preterm sex, missing
```

Don't forget to save typing by recalling commands, either by clicking on them in the review window or by using the Page-Up key.

1.4 Tables of means and other things

The command `tabulate` is only useful for making frequency tables. The entries in the cells of the table are the frequencies of the values taken by a categorical variable, or by the combinations of two categorical variables. To make more general tables the command `table` is used. This command can also be used to make frequency tables, but many other things can be placed in the cells of the table. For example, to obtain the mean and standard deviation of `bweight` separately by `sex`, as well as the frequency of the `sex` categories, try

. `table sex, contents(freq mean bweight sd bweight)`

The first variable following the command determines the rows of the table. The word `contents` refers to the contents of the cells of the table and is usually abbreviated to c. The words `freq`, `mean`, and `sd` are among the keywords which are recognized by the `table` command, so the contents of the table in this case will be the frequencies, the mean birth weight, and the standard deviation of birth weight. To make a table of the median and the lower and upper quartiles for birth weight, by sex, try

. `table sex, c(freq median bweight p25 bweight p75 bweight)`

When the variable summarized in the cells of the table has missing values it can be useful to know how many non-missing values were used. The keyword `count` allows you to find this out. Try

. `table sex, c(freq count gestwks median gestwks)`

Here `freq` refers to the non-missing observations of `sex` while `count` refers to the non-missing observations of `gestwks`.

To make a two-way table showing the mean birth weight by both `sex` and `hyp`, try

. `table sex hyp, c(freq mean bweight)`

The first variable following the command is `sex`, and this determines the rows of the table. The second variable (`hyp`) determines the columns.

The command `tabstat` is very useful for summarizing several numeric variables in the same table. For example, to produce a table showing the mean and standard deviation of birth weight, weeks of gestation and maternal age, try

. `tabstat bweight gestwks matage, statistics(mean sd)`

where `mean` and `sd` are two of several keywords that go with the option `statistics`. When the option `statistics` is not used mean values are given by default. If you want them separately by sex try,

. `tabstat bweight gestwks matage, statistics(mean sd) by(sex)`

1.5   Restricting the scope of commands

We have seen that Stata commands can be restricted to observations $1, 2, \ldots, 5$ (for example), by adding `in 1/5` to the command. Commands can also be restricted to operate only on observations which satisfy given conditions. The conditions are added to the command using `if` followed by a logical expression.

An example of a logical expression is 'birth weight less than 2000 g'. This can only be true or false. The expression would be false for a subject with a birth weight equal to 3000 g. To restrict the command `list` to observations for which this condition is true, try

```
. list bweight if bweight < 2000
```

If the logical expression `bweight < 2000` is true the observation is listed, but not otherwise. Other useful logical expressions are `X == Y` for equal, `X <= Y` for less or equal, `X >= Y` for greater or equal, and `X != Y` or `X ~= Y`, for not equal. A common error is to use = in a logical expression instead of ==. This is wrong because `X = 1` asks Stata to assign the value 1 to the variable X while `X == 1` is a logical expression, which does not change the value of `X`.

A command that uses logical expressions is `count`. It is particularly useful when exploring data. For example

```
. count if bweight <= 2000 & sex == 1
```

counts the number of observations which satisfy both the logical expressions `bweight <= 2000` and `sex == 1`. Note that the symbol for logical and in Stata is &. Similarly,

```
. count if bweight <= 2000 | bweight > 4000
```

will count the number of babies whose birth weight was either less than or equal to 2000 g or greater than 4000 g. The symbol for logical or in Stata is |, which is usually at the left end of the bottom row of letters on your keyboard. It is indicated by a broken vertical line on the key.

1.6   Generating new variables

New variables are generated using the command `generate`. Try

```
. generate num1 = 1
. generate num2 = 2
. browse
```

The new variable `num1` takes the value 1 for all observations, while `num2` takes the value 2. New variables which are made up from old variables can also be produced with `generate`, together with the usual mathematical operations and functions:

$$+ \quad - \quad * \quad / \quad \ln \quad \exp \quad \hat{} \quad \text{sqrt}$$

The sign ^ means 'to the power of', ln means natural logarithm, and sqrt means square root. Some examples are

```
. generate num3 = num1 + num2
. generate num4 = num1/num2
. generate logbw = ln(bweight)
. browse
```

The variable `logbw` is the natural logarithm of birth weight. Note that `ln` could have been replaced by `log` as Stata treats them as synonyms. Logs to base 10 are obtained with `log10( )`. To browse `bweight` and `logbw` on their own try

```
. browse bweight logbw
```

To change the values of `num1` the command `replace` is used, as in

```
. replace num1 = 7
. browse num1
```

or

```
. replace num1 = 35 if hyp == 1
. browse num1 hyp
```

1.7 Ordering, dropping and keeping

The command

```
. describe
```

shows the list of variables in their current order. To change this order so that `id` and `sex` come first, try

```
. order id sex
. describe
. browse
```

You will see that the order has changed in the spreadsheet as well as the variables window. Variables that are no longer useful can be dropped, e.g.

```
. drop num1 num2
```

will drop the variables `num1` and `num2`. The command

```
. drop if sex == 1
```

will drop all records with `sex == 1`. The command `keep` does the opposite of `drop`, so that the command

```
. keep logbw
```

will drop all variables except for `logbw`. Once data are dropped there is no way of getting them back other than by re-loading the whole dataset with

```
. use births, clear
```

where `clear` gives permission for the memory to be cleared before the births data are re-loaded.

1.8   Sorting data

Stata can sort the records in a file according to values (numeric or string) of a variable. The file is not physically re-arranged – instead a key is created which tells Stata commands the order in which the records should be processed. Try the following:

```
. list id matage sex sexalph in 1/10
. sort matage
. list id matage sex sexalph in 1/10
```

The records are now sorted in ascending order of maternal age. To sort on `id` within `matage` try

```
. sort matage id
. list id matage sex sexalph in 1/10
```

You will see that within each value of `matage` the records are sorted in order of `id`. To sort on a string variable try

```
. sort sexalph
. list id matage sex sexalph
```

Note that `female` comes before `male` because `f` comes before `m` in the alphabet. To restore the original sort order, try

```
. sort id
```

1.9   Using Stata as a calculator

The `display` command can be used to carry out simple calculations. For example, the command

```
. display 2+2
```

will display the answer 4, and

```
. display 2^3
```

will display the answer 8. The command

```
. display ln(10)
```

will display the natural logarithm of 10, which is 2.3026, and

```
. display sqrt(25)
```

will display the square root of 25. Text can also be displayed as in

```
. display "The natural logarithm of 10 is  " ln(10)
```

Note that because the text contains spaces it must be surrounded by the quotes symbol " which is usually found above the number 2. The result can also be color-coded as in

```
. display as text "Square root of 25 is " as result sqrt(25)
```

The keywords `as text` and `as result` determine the colors: when the background is black `as text` displays are green and `as result` displays are yellow. Other display styles are `as input` (white) and `as error` (red).

Standard probability functions are readily available. For example, the probability below 1.96 in a standard normal (i.e. Gaussian) distribution is obtained with

```
. display norm(1.96)
```

while

```
. display 1 - norm(1.96)
```

will display the probability above 1.96. Similarly,

```
. display chi2(1,3.84)
```

will display the probability below 3.84 in a chi-squared distribution on 1 degree of freedom, and

```
. display chi2tail(1,3.84)
```

will display the probability above 3.84. Try `help probfun` for a full list of available probability functions.

1.10   Shortcuts

Variable names can be abbreviated, as long as the abbreviation is unique. Try

```
. list id matage hyp gestwks in 1/10
. list id mat hy gest in 1/10
. list i m h g in 1/10
```

Also lists of variable names can be shortened if they are consecutive

```
. list id-gestwks in 1/10
. list i-g in 1/10
```

or if they share some unique initial letters:

```
. list se* in 1/10
```

where se* stands for "all variables with names starting with se".

Command names, as well as options within commands, can be abbreviated (with a few exceptions). Try

```
. sum matage
. l   matage in 1/10
. br matage
. tab sex
```

Note that tab is accepted as an abbreviation for tabulate, *not* for table, which must be typed in full.

1.11   Stata syntax

The word syntax here refers to the rules which govern how a Stata command is put together. The heart of any Stata command takes the form

command   varlist   if_expression   in_range,   options

For example, try

```
. list bweight hyp if sex==1 in 1/10, table noobs
```

The *command* is list, the *varlist* is bweight hyp, the *if_expression* is if sex==1, the *in_range* is in 1/10, and the *options* are table noobs. Adding weights to a Stata command will be covered in Chapter 7.

1.12   Using the Stata help facilities

At the end of the top line on the Stata screen you will see the *Help* tab. Click on this and a small menu will appear. Click on *Stata Command* and you will see another menu in which the name of the command should be entered. For example, enter list and press *OK*. Alternatively just type

```
. help list
```

Each Stata command has a help file, but the amount of information can be rather overwhelming. One useful bit to look at is the line which shows the syntax:

```
list [varlist] [if exp] [in range] [, options ]
```

Parts of the syntax which are not essential are shown inside square brackets [ ]. In Stata commands the options are always entered after a comma. The syntax for `list` shows that there are several options available. If you scroll down the help screen you will see that they are described under the heading *Options*. Looking further down at the *Examples* section, you find some of the common ways in which the command is used.

Another sort of help available from the Stata screen is help on commands linked to particular operations. Cancel the current help screen, click on *Help*, and then on *Search*. Enter the keyword *tables* to search on, and you will see a long list of Stata commands relevant to making tables. These start with official commands from Stata and then go on to contributed commands which have been written by users and published in the *Stata Journal* (previously the *Stata Technical Bulletin*). We shall return to contributed commands in Chapter 20.

Exercises

1. Load the births data.

2. List the variables `bweight` and `hyp` for observations 20–25 inclusive.

3. Summarize all variables.

4. Summarize `matage` in detail.

5. Use `codebook` to find out more about `sexalph`.

6. Use `count` to find out how many hypertensive women there are.

7. Summarize `matage` for hypertensive women.

8. Use `count` to find how many hypertensive women have babies with birth weight less than 2000 g.

9. Use `count` to find how many women over 30 are hypertensive.

10. Tabulate the values of `sex`.

11. Make a table of mean birth weight by `sex`.

12. Make a table of median birth weight by `sex`.

13. Generate a new variable called `bwkgs` which is the birth weight in kilograms.

14. Use `display` to calculate $\sqrt{3^2 + 4^2}$.

15. Use `display` to find the probability above 4.3 in a chi-squared distribution on 1 degree of freedom.

Answers to these exercises can be obtained by running the program `chap1` with the command

. chap1

To get the answer for question 8 only, try

. chap1, q(8)

Chapter 2 has no answers, but answers for chapter 3 can be obtained with

. chap3

and similarly for the other chapters.

Chapter 2

Tabs, menus and dialog boxes

In Chapter 1 we showed how to do some basic things by typing commands into the command window. Stata 8 introduced dialog boxes for helping the user to create these commands, and we shall now demonstrate these by repeating some of the contents of Chapter 1.

2.1   Where to find the dialog boxes

At the top of the Stata screen you will see the tabs

   File   Edit   Prefs   Data   Graphics   Statistics   User   Window   Help

Selecting *Data*, *Graphics*, or *Statistics* produces one or more menus and selecting one of the choices in the menus will produce a dialog box which is used for entering information.

2.2   A first look at the data

To load the births data, select the *File* tab (top left hand of screen), then *Open*, and you will see the list of Stata data files in the working directory. Depending on how your version of MS-Windows is set up you will see either `births.dta` or `births` in the list (the extension `.dta` means the file is a Stata dataset). Now highlight `births.dta` (or `births`) and select *Open* to load the data. The Stata command which did this is

```
. use "C:\data\hs8\births.dta"
```

and is shown in both the results window and the review window. Note that the full path to the data file is shown, so this command will work whatever your working directory, unlike

```
. use births.dta
```

which requires the file `births.dta` to be in the working directory.

To describe the variables in the births dataset select the *Data* tab, then *Describe data*, then *Describe variables in memory*. This will bring up a dialog box with various options:

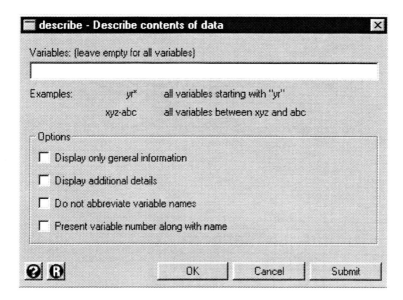

Ignoring the options and pressing *OK* produces and executes the command

```
. describe
```

Pressing *Submit* does the same, but leaves the dialog box on the screen. Provided you know the command name you can bring up the dialog box directly by typing the command

```
. db describe
```

where `db` stands for dialog box. This can be helpful when you know the command name but have forgotten the options. Dialog boxes remember the entries from a previous use, so it may be necessary to press R, at the bottom left of the dialog box, to clear any previous selections. Pressing ? (next to R) brings up the help file for the command. Pressing *Cancel* closes the dialog box.

To obtain a summary of the data, select the *Statistics* tab, then *Summaries tables & tests*, then *Summary statistics*, then *Summary statistics* which calls up the dialog box. Simple summaries are the default here, so press *OK* to produce the command

```
. summarize
```

Alternatively you can cut out the menus and use

```
. db summarize
```

to go straight to the dialog box. For a more detailed summary of the variable `gestwks` select the second option, then enter `gestwks` in the *Variables* box by placing the cursor in the box and clicking on `gestwks` in the Stata variables window. Pressing *OK* will produce the command

```
. summarize gestwks, detail
```

Names can also be typed directly, and unique abbreviations can be used.

To list the values of `matage`, select the *Data* tab, then *Describe data*, then *List data*. Enter `matage` in the *Variables* box, and press *OK* to produce the command

```
. list matage, separator(5)
```

Click on the *Break* icon to stop the listing. The option `separator(5)` refers to the number of rows printed before a separator line is drawn, and is the default. It can be changed in the *Options* tab in the section called *Table options*. To suppress the listing of observation numbers check the box at the bottom of the *Main* tab of the dialog box to produce the command:

```
. list matage, noobs separator(5)
```

Stata commands can be restricted to observations 1, 2, ..., 5 (for example) by using the *by/if/in* tab in a dialog box. Try

```
. db list
```

to bring up the dialog box, and press R, at the bottom left of the dialog box, to clear previous selections. Enter `matage` in the *Variables* box, select the *by/if/in* tab, check *Obs. in range*, and fill in 1 to 5 (you can type 5 or use the spinner). Press *OK* to produce the command

```
. list matage in 1/5, separator(5)
```

Leaving the *Variables* box blank will list the data for all the variables in the dataset.

2.3 Tables of frequencies

To obtain the frequencies of the values taken by the categorical variable `hyp`, select the *Statistics* tab, then *Summaries, tables & tests*, then *Tables*, then *One-way tables*. Enter `hyp` in the *Categorical variable* box, and press *OK* to produce the command

```
. tabulate hyp
```

To obtain the two-way frequency table of `hyp` and `sex`, select the *Statistics* tab, then *Summaries, tables & tests*, then *Tables*, then *Two-way tables with measures of interaction*. Enter `hyp` in the *Row variable* box and `sex` in the *Column variable* box, and press *Submit* to produce the command

```
. tabulate hyp sex
```

The basic output from a cross-tabulation reports frequencies only; to include relative frequencies for rows or columns check the option *Within–row relative frequencies*, or the option *Within–column relative frequencies*, in the *Main* tab of the dialog box. These options produce the commands

```
. tabulate hyp sex, row
. tabulate hyp sex, column
```

2.4   Tables of means and other things

To make a table showing the mean and standard deviation of `bweight` by `sex`, together with frequencies, select the *Statistics* tab, then *Summaries tables & tests*, then *Tables*, then *Table of summary statistics (table)*. Enter `sex` as the row variable, and elect *Frequency* for the first statistic. Select *Mean* for the second statistic and enter `bweight` for the corresponding variable. Select *Standard deviation* for the third statistic and enter `bweight` for the corresponding variable. Press *OK* to produce the command

```
. table sex, contents(freq mean bweight sd bweight)
```

The frequencies refer to the two categories of `sex`. To add the number of non-missing values for `bweight` in each category of sex, select *Count non-missing* as the fourth statistic and `bweight` as the corresponding variable. To make a table of the median and the lower and upper quartiles for birth weight, by sex, start with

```
. db table
```

to bring up the dialog box, and press R to clear the previous selections. Then enter `sex` as the Row variable and select *Frequency* for the first statistic. Select *Median* for the second statistic and enter `bweight` for the corresponding variable. Select *Percentile* for the third statistic, use the spinner to select 25, and enter `bweight` for the corresponding variable. Select *Percentile* for the fourth statistic, use the spinner to select 75, and enter `bweight` for the corresponding variable. Press *OK* to produce the command

```
. table sex, c(freq median bweight p25 bweight p75 bweight)
```

To make a two-way table showing the mean birth weight by both `sex` and `hyp`, start with

```
. db table
```

and press R to clear the previous selections. Then enter `sex` as the *Row variable*, check the *Column variable* box, and `hyp` as the *Column variable*. Select *Frequencies* as the first statistic. Select *Mean* as the second statistic with `bweight` as the corresponding variable. Press *OK* to produce the command

```
. table sex hyp, c(freq mean bweight)
```

Another way of making a table showing the mean and standard deviation of several metric variables, such as `bweight`, `gestwks` and `matage`, separately by `sex` is to select the *Statistics tab*, then *Summaries tables & tests*, then *Tables*, then *Table of summary statistics (tabstat)*. Enter `bweight`, `gestwks` and `matage` in the Variables box. Select *Mean* for the first statistic. Select *Standard deviation* for the second statistic. Check *Group statistics by variable* and enter `sex` in the accompanying box. Press *OK* to produce the command (all one line on the screen)

```
. tabstat bweight gestwks matage, statistics(mean sd)
        by(sex) columns(variables)
```

2.5   Restricting the scope of commands

To list the `id` and `bweight` for those observations for which the logical condition 'Birth weight less than 2000 g' is true, start with

```
. db list
```

to bring up the dialog box. Then press R, enter `id` and `bweight` in the *Variables* box, and select the *by/if/in* tab, enter `bweight < 2000` in the *if* box, and press *OK* to produce the command

```
. list id bweight if bweight < 2000, separator(5)
```

2.6   Generating new variables

To generate a new variable called `num1` which takes the value 1 for all records, select the *Data tab*, then *Create or change variable*, then *Create new variable*. Enter `num1` in the *Generate variable* box, and 1 in the *Contents* box. Select *byte* in the *generate variable as type* menu, and press *OK* to produce

```
. generate byte num1 = 1
```

The keyword `byte` refers to how the variable will be stored (see Chapter 3). To generate a new variable, `logbw`, which is the natural logarithm of birth weight, try

```
. db generate
```

to bring up the dialog box. Then press R, enter `logbw` in the *Generate variable* box, select *Create*, double click ln(), position the cursor between the parentheses ( ) and enter `bweight`. Press *OK* to produce

```
. generate float logbw = ln(bweight)
```

To change the values of `num1` to 7 (for example), select the *Data* tab, then *Create or change variables*, then *Change contents of a variable*. Enter `num1` in the *Variable* box, and 7 in the *Contents* box. Press *OK* to produce

```
. replace num1 = 7
```

2.7   Ordering, dropping and keeping

The command

```
. describe
```

shows the list of variables in their current order. To change this order so that `id` and `sex` come first, select the *Data tab*, then *Variable utilities*, then *Change order of variables in the dataset*. Enter `id` and `sex` in the box labelled *Variables to move to the front*, and press *OK* to produce

```
. order id sex
```

To drop the variables `logbw` and `num1`, select the *Data* tab, then *Variable utilities*, then *Eliminate variables or observations*. Check both *Drop* and *Drop variables*, enter `logbw` and `num1` in the *Drop variables* box, and press *OK* to produce

```
. drop logbw num1
```

These two variables will then be dropped regardless of any if/in conditions which you might enter using the *by/if/in* tab. To drop observations which satisfy some logical condition check *Drop* but not *Drop variables*: instead fill in the condition using the *by/if/in* tab.

2.8   Sorting data

To sort the data according to the value of `matage` (lowest first), select the *Data* tab, then *Sort*, then *Sort data*. Enter `matage` in the *Variables* box, and press *OK* to produce

```
. sort matage
```

To sort on `id` within `matage` enter the variables `matage` and `id` in the *Variables* box, in that order. The box labelled *Perform stable sort* ensures that tied observations are left in the original order within any ties.

2.9   Using Stata as a calculator

To calculate $2 + 3$, select the *Data* tab, then *Other utilities*, then *Hand calculator*. Press *Create*, use the arithmetic keypad to build the expression $2 + 3$, and press *OK*. Press *OK* (again) to produce

```
. display 2 + 3
```

To build the expression ln(10) try

```
. db display
```

to bring up the dialog box, and press R. Then press *Create*, double click ln() to put $\ln(x)$ in the box at the top, use the keypad to replace x by 10, and press *OK*. Press *OK* (again) to produce

```
. display ln(10)
```

Standard probability functions are readily available. For example, to obtain the probability below 1.96 in a standard normal (i.e. Gaussian) distribution, try

```
. db display
```

to bring up the dialog box and press R. Then select *Create*, select *Probability*, scroll down for norm(), double click norm(), use the keypad to build the expression norm(1.96), and press *OK*. Press *OK* (again) to produce

```
. display norm(1.96)
```

Similarly the probability below 3.84 in a chi-squared distribution on 1 degree of freedom, is found by selecting chi2() and building the expression chi2(1 , 3.84).

As a result of repeating some of Chapter 1 using the menus and dialog boxes, you will have seen (we hope) that typing commands is simpler and quicker than using menus and dialog boxes. Menus can be useful when you don't know the name of the command to carry out your requirements, and dialog boxes can be helpful when selecting options, particularly for complex commands.

Exercises

Try some of the exercises in Chapter 1 using menus and dialog boxes rather than commands. There are no solutions for this chapter.

Chapter 3

Housekeeping

Housekeeping refers to all those small jobs which are a nuisance at the time, but make life easier later. This chapter covers how to label and add notes to datasets; how to label variables and their values; how to recode variables and deal with codes for missing values; how to manage dates; how to save datasets; and how to use log and batch files.

3.1   Labelling a dataset

A label can be attached to a dataset to remind you what the data refer to, and when the dataset was formed. The births dataset has already been labelled. After

```
. use births, clear
```

you will see the label 'Data from 500 births' within parentheses. To change this label (or create one) try

```
. label data "Whatever you like"
. describe
```

The label appears in the output of `describe` below the file name and above the date and time when it was last saved. Note that when the label includes spaces, you must enclose it in quotes " ". The quotes are inserted automatically when you use the dialog box to label a dataset, so don't put them in yourself.

3.2   Notes

You can also add notes to a dataset as a reminder of things you have done or should do in the future. For example

```
. notes: This dataset was created by George.
```

The text in this command does not require quotes. A time stamp (TS) can also be inserted with

```
. notes: TS This dataset was created by George.
```

and the notes can be displayed with

```
. notes list _dta
```

Notes can be specific to variables, and this can be a useful way of reminding yourself how variables are coded. Several such notes have been attached to the births dataset, and can be displayed with

```
. notes list
```

to show all the notes, or

```
. notes list hyp
```

to show only the notes for hyp. To attach a note to a variable, try

```
. notes matage: maternal age in years
```

See help notes for how to drop notes. Any notes you create will be lost when you clear the data from memory; to make them a permanent part of the dataset you will need to save the new version of the data (see Section 3.8).

3.3 Labelling variables and their values

After

```
. describe
```

you will see the variable names on the left and their labels on the right. A variable label is used to record further information about the variable. Now try

```
. label var gestwks "blah blah blah"
. describe
```

The new label for gestwks has replaced the old one. To remove a label simply label the variable with nothing, as in

```
. label var gestwks
. describe
```

Any labels you create will be lost when you clear the data from memory; to make
them a permanent part of the dataset you will need to save the new version of the
data (see Section 3.8).

The values which a categorical variable takes can also be labelled. It is nearly
always a good idea to label variables, but not necessarily a good idea to label their
values. By default the labels appear instead of the values in the output of commands
such as `tabulate`, which can sometimes be a nuisance. In addition, if values are
re–coded for any reason, you have to remember to update the labels accordingly. On
the other hand, labelling values means you don't have to remember what the values
stand for.

Labelling variable values involves setting up a named mapping between the values
and their labels. For example, to label the values of `hyp` using the mapping `hypmap`
in which $1 \rightarrow$ hyper, $0 \rightarrow$ normal, try

```
. label define hypmap 1  "hyper"  0  "normal"
. label values hyp hypmap
```

To check the labels use

```
. label list hypmap
```

The frequency table of `hyp` will now show the label for each of its values

```
. tabulate hyp
```

To see the results of `tabulate` without the labels, or to browse the numerical values
of this variable without the labels, use the option `nolabel`:

```
. tabulate hyp, nolabel
. browse hyp, nolabel
```

The same label mapping can be used for several variables, if appropriate. To stop
using `hypmap` to label the values of `hyp`, try

```
. label values hyp
. tabulate hyp
```

which labels the values of `hyp` with nothing. To drop `hypmap` altogether, try

```
. label drop hypmap
```

3.4   Data types and display formats

There are five data types in Stata: byte, integer, long integer, float, and double. What
distinguishes them is how much memory they take to store. Variables of type byte
take a single byte per observation; integer variables take 2 bytes; long integers take
4 bytes, floating point variables take 4 bytes, and double precision variables take 8
bytes. If you try

```
. describe
```

you will see that `bweight` is stored as type float, while `lowbw` is stored as type byte. These different types need not affect the user - their main function is to economize on memory.

Next to the type, in the results of `describe`, you will see the display format. The default display is a number of width 8, 9, or 10 digits, depending on type. This default works perfectly well without further thought from the user, but it can sometimes be useful to change it, using the `format` command. For example, when listing variables in columns, you may wish to format them to take less room. The Stata symbol for format is % and `%5.2f` means that numeric variable values should be displayed in the form `xx.xx` (i.e. with total width 5, 2 of which are after the decimal point). For string variables `f` is replaced by `s`, and the format `%10s` means that 10 is the maximum length of the string, while `%-10s` will left-justify the string. Try

```
. list gestwks in 1/5
. format %4.1f gestkws
. list gestwks in 1/5
```

and see the difference. Note that if the width of a `%f` display format is not adequate to display the data, Stata will over-ride this. See section 15 of the User's Guide for more details about data storage and display.

3.5   Recoding a variable

Variables can be recoded using `recode`. For example, to create a new variable `sex2` which is the same as `sex` but coded 1 for male and 0 for female, try

```
. recode sex 2=0, generate(sex2)
```

You can check the recoding with

```
. tabulate sex2 sex
```

Of course we could have recoded the variable `sex` directly, but it is generally better to create a new variable rather than to change the original one.

3.6   Missing values

The missing numeric value symbol in Stata, up until version 7, was an isolated decimal point. Stata 8 added 26 additional symbols, namely

$$.a \quad .b \quad .c \quad \cdots \quad .z$$

which can be useful when it is necessary to distinguish between reasons why the values are missing. When making comparisons or sorting, the following rules are observed:

- all numbers are less than .

- . is less than .a

- .a is less than .b

- .b is less than .c , and so on up to .z

In the births data there are 10 missing values for the variable `gestwks`. With most commands Stata automatically excludes records with missing values in any of the variables mentioned in the command. For example

```
. summarize gestwks
```

shows a summary based on the 490 records with non-missing data.

Particular care is needed when using > with missing values, because all missing values are larger than any number. For example, `gestwks` is missing for 10 subjects, but

```
. count if gestwks > 15
```

returns 500, not 490, because . is > 15. To avoid including the missing records in this count you need to exclude them yourself with

```
. count if gestwks > 15 & gestwks < .
```

In some datasets missing values are identified by a code like 9 or −1. To make sure Stata recognizes such values as missing you should change them into a missing value symbol with `mvdecode`. The data file `births_miss.dta` has the missing values coded as −1. Load these data with

```
. use births_miss,clear
. summarize
```

and change all occurrences of −1 into the missing code `.a` with

```
. mvdecode _all, mv(-1=.a)
. summarize
```

3.7   Dates

Dates are tricky to deal with because they are usually coded as string variables, but in order to be able to compare dates, or to calculate elapsed time, it is necessary to convert them into time since some fixed date. Stata uses the convention that dates are coded as days since 1/1/1960, so dates before 1/1/1960 are negative numbers, and dates after are positive numbers.

The function `mdy()` returns the number of days since 1 Jan 1960. For example

```
. display mdy(1,1,1960)
```

returns 0, and

```
. display mdy(1,31,1960)
```

returns 30. The first number between the brackets indicates the month, the second the day and the last the year (respectively the m, the d and the y of mdy). Check how many days there are between 4 Sep 2000 and 1 Jan 1960. Make sure you don't leave a space between mdy and ( ) and make sure that you enter 4 Sep 2000 as (9,4,2000) not (4,9,2000). The answer is 14,857 days.

The function date() does the same thing for dates which are held as string variables. Several formats are allowed. For example

```
. display date("31/jan/1960", "dmy")
. display date("31/1/1960", "dmy")
. display date("31-jan-1960", "dmy")
. display date("31jan1960", "dmy")
```

all give the same results. The second argument in the date function is used to state the order in which the days, months and years appear. Try

```
. display date("jan/31/1960", "mdy")
```

For a full list of available date functions, try

```
. help datefun
```

In practice most datasets hold dates in the form dd/mm/yyyy (Europe) or mm/dd/yyyy (USA). A simple example is shown in the file dates. Load these data in memory and describe them with

```
. use dates, clear
. describe
. codebook start
```

You will see that start is a string variable which records dates in the European form (dd/mm/yyyy). To generate a new variable datein with dates in Stata form, try

```
. generate datein = date(start, "dmy")
. list start datein
```

Note that the new variable datein contains days since 1/1/1960, and is numeric. To get the best of both worlds (alphabetic and numeric) we can format the numeric variable datein so that it is displayed as a date, using

```
. format datein %d
. list start datein
```

See the *Stata User's Guide* and the Stata FAQs (Frequently Asked Questions) to find out how to deal with dates in which the century has been omitted, such as 21/9/76, and dates without separators, such as 21091976.

3.8   Saving files

It is sometimes necessary to save the data in memory, for example when labels and notes have been added. This is done with the command `save` which creates a new file on disk, in Stata format, containing the data which are currently in memory. By default, the new file is given the extension `.dta`, and it is saved in your working directory. To over-write a file which already exists in this directory requires the option `replace`. Load the births data and save them in a new file called `mybirths` with the command

```
. use births, clear
. save mybirths
```

Now repeat the command

```
. save mybirths
```

and Stata will refuse to do it, but

```
. save mybirths, replace
```

will work. Datasets saved with `save` in Stata 8 cannot be read by Stata 7 - the command to save data in Stata 7 format is `saveold`.

3.9   Log files

To keep a record of the results obtained while using Stata you can open a log file by clicking on the *Log* icon at the top of the screen (the one with a rolled up document and a traffic light). If the log file is a new one you will be asked to name it: choose, for example, the name `house`. By default the log file will be saved in your working directory with the name `house.smcl`. The extension `.smcl` stands for Stata markup and control language.

If `house.smcl` already exists in your working directory, Stata will ask whether to append the new results to the existing file or to overwrite it. Once the log is open, run a few commands such as

```
. use births, clear
. describe
. summarize
. tabulate hyp
```

To look at the log file while you are still working in Stata you can click on the *Log* button again and select *View snapshot*. If you keep this window open while you work you will need to *Refresh* it to view your latest results. Otherwise open and close it as you go along. You can copy and paste from this window into your favourite wordprocessing program, or print it as it stands. To close a log file click on the *Log* icon and select *Close* log file.

Log files record both commands and output. Their main purpose is to enable the user to import parts of the output into a wordprocessor for eventual printing and publication. Of course if you keep a log file open for the whole of a session it will contain a long record of everything that happened during the session. This is not a very efficient way of working. Much better is first to organize your work in a batch file.

3.10   Do files

A do file, or batch file, is a text file which contains a list of Stata commands. Its name should have the extension `.do`. It can be created using the *Do-file Editor*, which you open by clicking the icon at the top of the screen that looks like an envelope. Once you are in the Do-file Editor enter the commands

```
use births, clear
describe
summarize
tabulate hyp
```

Save this do file in your working directory with the name `myhouse.do` using the Do-file Editor's menu. Then, in the command window, you can execute it with the command

```
. do myhouse
```

If Stata does not find your do file you have probably saved it in a directory which is not your working directory - go back to the Do-file Editor and check this. If Stata does find your do file it will obey the commands in the file.

A sensible way of working is to open a do file at the start of a session, and to add commands to it as you go along. Each time you run the do file Stata will obey all the commands in it. Comments can be included in the file, but each comment line must start with an asterisk (*). If you want to leave out one of the commands in your do file, convert it into a comment by inserting * at the beginning. Once you are satisfied that your set of commands is complete, you can open a log file and run the do file one last time to keep a log of the output.

There is another advantage of using do files: they can be re-run whenever you wish and provide a record of what you have done. Because of this it is not necessary to create different versions of the data as new variables are generated and recoded. Instead the generating and recoding can be organized in do files which are run before the analysis starts, and the original data file is left unchanged.

Exercises

1. Open a log file. Load the births data and describe the variables. Replace the current label for the variable `hyp` with the new label "Maternal hypertension", and use `describe` to check that your labelling has worked.

2. Use the command `label define` to create a label mapping called `preterm` which maps the 0 to "normal" and 1 to "pre-term". Use `label list` to check that you have done this correctly.

3. Use the label mapping `preterm` to label the values of the variable `preterm`. Check the results by tabulating `preterm` with and without the option `nolabel`.

4. Create a do file with the *Do-file Editor* (think of a name). Include in this do file the commands to load the births data, label the values of the variables `preterm` and to tabulate `preterm`. Run the do file, and use `browse` to check that the do file has done what you wanted it to do.

5. Use the `date()` function to calculate the number of days which have elapsed between 1/jan/1960 and 4/sep/2000. Do the same for 2000/sep/4.

6. Load the data in the file `dates.dta` and use the `date` function to create two new numerical variables `datein` and `dateout` from the string variables `start` and `stop`. List all the variables. Use the `format` command to format the new variables as dates, and list all the variables again.

7. Calculate the number of days between `datein` and `dateout` for each of the 6 subjects by creating a new variable `days`. List its values and then generate the equivalent number of years by dividing `days` by 365.25.

8. Inspect the contents of the log file by clicking on the Log icon, and checking *View snapshot*. To close the log file, click the Log icon again and check *Close log file*.

Chapter 4

Data input and output

This chapter covers how to load data from text files which have been created by spreadsheets, wordprocessors or text editors; how to load data directly from the keyboard; and how to save Stata data as text files.

4.1 Data sources

Datasets which are stored in Stata format are binary files with extension `.dta`. For example, the births data are in the Stata file `births.dta` which can be loaded into memory with the command `use births`. The command `type` can be used to type out the contents of a file without loading it into memory. Try

```
. type births.dta
```

and you will see gobbledygook – this is because Stata data files are binary.

To get data into Stata format it is necessary to start from a file of some sort, load this into memory, and then save it as a Stata file with the command `save`. Stata is capable of loading data which have been stored in almost any form, but we shall concentrate on the two most common forms:

- Text files in which the values of the variables are separated by the tab or comma symbol; these are commonly obtained from spreadsheet programs such as Excel, or database programs such as Access.

- Text files in which the values of the variables are separated by the space symbol; these are commonly obtained from wordprocessors, such as Word, or from a text editor.

In both cases the names of the variables come first, followed by the values. To illustrate the methods we have prepared files containing the births data in both ways:

`births.tab` is a text file which is tab-separated, and `births.dct` is a text file which is space-separated. The reason for the extension `.dct` will become clear in a moment. To look at the contents of `births.tab` try

`. type births.tab`

The first line contains the names of the variables, and the following lines contain their values. Click on the *Break* icon (the red ⊗ icon at the top of the screen) to stop scrolling. To check that the values really are separated by the tab symbol, try

`. type births.tab, showtabs`

To look at the contents of `births.dct` try

`. type births.dct`

and to make sure that the values are *not* separated by tabs, try

`. type births.dct, showtabs`

Note that in this file the variable names are placed in a dictionary (i.e. between curly brackets and preceded by the word `dictionary`), which is why the file has the extension `.dct`. This will be discussed further in Section 4.5.

4.2   Data from a spreadsheet

When the original data are kept in a spreadsheet, the first step – before trying to read them into Stata – is to save them as a text file in which the values of the variables are separated by a tab (or comma) symbol. There are three things to watch out for when doing this:

1. Make sure that the variable names do not include spaces – you may need to change the spaces into something like underscore before saving the text file, if they do.

2. Make sure that missing values are coded with a blank (*not* a space).

3. Many spreadsheets and database packages will automatically save text files with the extension `.txt`. If you want to save a text file with the extension `.tab` you must enclose the file name in quotes, e.g. `"births.tab"`.

The births data were originally stored in Excel as `births.xls`. To create the Stata file `births.dta` you would open `births.xls` within Excel and save it as a tab-separated text file with the name `"births.tab"`. The file `births.tab` is then loaded with the command `insheet`, as follows:

`. insheet using births.tab, clear`
`. describe`

At this stage you would normally save the data in memory so that in future it can be loaded as a Stata data file. Don't do it in this case, because you already have `births.dta` on disk. When the values of a text file are separated with a comma, the option `comma` is used with the command `insheet`. There is also an option `delimiter("X")` which can be used to set X, or any other character, as the delimiter.

String variables cause no problem with tab- or comma-separated files. The command `insheet` will recognize when a variable takes strings as values, and act accordingly. Sometimes, however, when spaces are entered in an Excel cell which should have been left blank, `insheet` may treat the corresponding variable as a string variable even though all the other values are numerical. The FAQ: *How do I get information from Excel into Stata* on the Stata website (*www.stata.com*) shows what to do when this happens.

4.3 Data from a wordprocessor

When the original data are kept in a wordprocessor, the first step is to save them as a text file in which the values of the variables are separated by a space symbol. Again there are three things to watch out for:

1. Make sure that the variable names do not include blanks.

2. Make sure that missing values are coded using a Stata missing value symbol.

3. Edit the file and enclose the variable names inside a dictionary, as with `births.dct`.

The command `infile` is used to load the data from `births.dct`, as in

```
. infile using births.dct, clear
. describe
```

Data on string variables can also be loaded with `infile` but each of them must be defined as such in the dictionary. This is done by writing `str#` in front of the variable name, where # is the maximum length of the string for that variable. Try

```
. type births.dct
```

to see how the string variable `sexalph` is defined.

4.4 Large datasets

Stata stores the whole of a dataset in memory, so the size of the dataset is limited by the size of the memory on your machine. Fortunately most computers these days are equipped with at least 256MB of memory, so this is rarely a problem. Be warned, however, that the default allocation of memory to Stata may be as small as 1MB. This can be increased to 16MB (for example) with

```
. clear
. set memory 16m
```

The `insheet` command will automatically choose the data type for the variables so that the dataset fits into the smallest possible amount of memory. To achieve the same with `infile` you need first to load the data, and then compress them. Try

```
. infile using births.dct, clear
. describe
```

and you will see that all the numeric variables have been stored as float. Now try

```
. compress
```

and you will see that `bweight`, for example, has been squeezed into an integer, which takes 2 bytes, although originally it was stored as a float, which takes 4 bytes.

With very large datasets it may be necessary to `infile` the data in parts, compress each part separately, and then append the compressed parts together again (Chapter 8 gives more details).

4.5   More about dictionary files

A dictionary can also contain information about how the variables are to be stored and also how to label them. Consider, for example, the dictionary file called `births2.dct`, which you can look at with

```
. type births2.dct
```

This contains instructions about the labels to be used for the variables and the type of storage required. When you load it with the command,

```
. infile using births2.dct, clear
. describe
```

you will see that the variables have been labelled and also that they have been stored as integer, float, byte, or string, depending on the storage type used in the dictionary. Note that the option `clear` can be used with both `insheet` and `infile`.

If you are faced with a file format which is outside the range of the methods described here, the *Stata User's Guide* (24.2) has an excellent step-by-step account of how to load data into memory, which covers all the possibilities.

4.6   Loading data from the keyboard

Only in rare cases would one load data directly into Stata from the keyboard, but it can be useful for very small datasets. It is best to do this with the *Data Editor*, after clearing any data from memory with

Table 4.1: Mortality rates per 100 000 by employment grade and categories of age

grade	agecat	rate
1	40	4.9
1	45	6.1
1	50	13.2
2	40	6.4
2	45	8.8
2	50	16.2

```
. clear
```

Start by clicking on the *Data Editor* icon, which is next to the *Data Browser* and has a sheet of paper with columns of entries.

To load the values of the 3 variables `grade`, `agecat` and `rate`, shown in Table 4.1, type their values in the spreadsheet that appears when the *Data Editor* is called. Don't type in the names of the variables as the first row - just type the values, column by column. After typing each value, press the ⏎ key. Stata automatically names each column (i.e. each variable) as `var1`, `var2`, etc. To change these names double click on the relevant column to open a dialog box and change `var1`, for example, into `grade`. When you have completed the data entry, close the *Data Editor* and check your editing by listing the data in memory. Any mistakes can be corrected by recalling the *Data Editor*.

4.7 Data output

Results which appear on the screen can be copied, and then pasted into another document. Simply highlight the area to be copied using the mouse, then click on *Edit* and select *Copy Text*. To copy and paste into a spreadsheet select *Copy Table*.

The ability to save the data in memory as a text file can also be useful. To see what is currently in memory, try

```
. describe
```

It should be the mortality rate data from Table 4.1. To save these data as a text file called `grade.raw`, without variable names, try

```
. outfile using grade.raw
. type grade.raw
```

To include a dictionary, type

```
. outfile using grade.dct, dictionary
. type grade.dct
```

The command `outfile` is the output equivalent of `infile`. The output equivalent of `insheet` is `outsheet`, and can be used to output text files with tab separators. For example, try

```
. outsheet using grade.tab
. type grade.tab, showtabs
```

The commands `save`, `outsheet` and `outfile` all allow the option `replace`.

4.8 Import and export to other packages

Stat/Transfer is the name of a general purpose program for moving data from one statistical package to another. It has nothing to do with Stata, but users of Stata have generally reported favourable experiences with it. See *www.stattransfer.com* for more details about this program.

If you wish to import/export to SAS XPORT transfer files, the Stata commands `fdause` and `fdasave` will be of interest. The `fda` refers to the US Food and Drug Administration, and its presence in the names of these commands is explained by the fact that they are mainly used for communicating with the FDA. See `help fdause` for further details.

The command `odbc` allows Stata to load, write, or view data from sources which support Open DataBase Connectivity, a standardized set of function calls that can be used to access data stored in both relational and nonrelational database management systems. Try `help odbc` for more information.

Exercises

1. Load the data in `example.dta` and have a look at the values of the variables with `list`.

2. These data are also in the tab-separated file `example.tab`. Use `type` to inspect this file, then load it into memory using `insheet`. Check whether these data correspond to those of `example.dta`.

3. The same data are also in the dictionary file `example.dct` which has space-separated values. Use `type` to inspect this file, use `infile` to load the data into memory, and check that the data which have been loaded are the same as the data in the file.

4. Clear the memory and use the Data editor to load the data shown below directly from the keyboard.

```
id    time    grp    sex
 6    21.7     2      f
 7     6.2     2      m
 8    35.4     1      m
 9    20.0     1      f
10     9.3     2      m
```

Close the Data editor, and list the data in memory.

Chapter 5

Graph commands

All the graph commands were upgraded in Stata 8. They now allow easy access to high-quality graphs plus the ability to over-ride the defaults and to arrange the layout in virtually any way you want. However, all this came at a price: there is no backward compatibility with the graphics commands in previous versions. It is still possible to access the older commands by typing `graph7` where you would have typed `graph` in Stata 7, but to access Stata 8 graphics you need to learn new commands, or call up the relevant dialog boxes using the *Graphics* menu. In this chapter we show how to produce basic graphics such as box plots, histograms, scatter plots, and line plots, using commands. In the next chapter we shall introduce the *Graphics* menu.

5.1 Box plots

To make a box plot of birth weight using the births data, try

```
. use births, clear
. graph box bweight
```

After a few seconds you will see a box plot where the box indicates the median and the two quartiles. The vertical lines above and below the box indicate the range of values, with outliers shown as separate points (see the *Stata Graphics Manual* for a precise definition of range and outlier). To make the box thinner, try

```
. graph box bweight, outergap(50)
```

which specifies that the gap between the edge of the graph and the first box should be 50% of the width of the box. The box can be drawn horizontally with

```
. graph hbox bweight, outergap(50)
```

The main use of box plots is when several are placed side by side, as in the comparison of the distribution of birth weight for boys and girls. Try

```
. graph box bweight, outergap(50) over(hyp)
. graph box bweight, outergap(50) over(hyp) over(sexalph)
```

The command

```
. graph box bweight, outergap(50) over(hyp) by(sexalph)
```

does almost the same thing, but by( ) makes two separate plots within the graph, while over( ) puts both boxes on one plot. The by( ) option for graph box can only be used with over( ).

5.2   Histograms

To make a histogram of birth weight, try

```
. graph twoway histogram bweight
```

The word graph makes it clear that twoway is a graph command, but it can be omitted, and we shall follow this practice in what follows. The rectangles in the histogram are called bins by Stata, and the height of each bin is the relative frequency per unit of birth weight (i.e. per gram). The appearance of the graph is not too bad, but you might wish to change where the first bin starts, how wide the bins are, and the scale used for the vertical axis from density to percent. Try

```
. twoway histogram bweight, percent start(0) width(500)
```

There is no over option with twoway histogram but by can be used. Try

```
. twoway histogram bweight, percent start(0) width(500) by(sexalph)
```

and

```
. twoway histogram bweight, percent start(0) width(500) by(sexalph hyp)
```

The command

```
. histogram bweight, percent start(0) width(500)
```

without the twoway does the same thing as twoway histogram, but you should be aware that they are different commands, although they share many options. In particular histogram allows you to superimpose a normal curve on the histogram, while twoway histogram does not. Try

```
. histogram bweight, percent start(0) width(500) normal
```

to see the superimposed normal curve which has the same mean and standard deviation as the variable bweight.

The command twoway histogram (and also histogram) can be used for categorical data with the option discrete. Try

Command	What it does
`graph twoway histogram varname`	Histogram
`graph twoway scatter yvar xvar`	Scatter plot
`graph twoway lfit yvar xvar`	Best fitting line
`graph twoway line yvar xvar`	Line plot
`graph twoway connected yvar xvar`	Connected point plot
`graph twoway rcap lower upper xvar`	Vertical capped lines for confidence intervals

Table 5.1: Commonly used twoway commands

```
. twoway histogram hyp, discrete
```

The appearance could be improved by using `xlabel` to label the X-axis with 0 and 1, by using `xscale` to extend the range of the X-axis to run from -1 to 2, and by using `gap` to include a gap between the bins, equal to 50% of the bin-width. Try

```
. twoway histogram hyp, discrete xlabel(0 1) xscale(range(-1 2)) gap(50)
```

The `twoway` command includes many possible plot-types, and some of the more commonly used ones are listed in Table 5.1.

5.3 Scatter plots

Scatter plots can be used to evaluate the association between two metric variables such as `bweight` and `gestwks`. Try

```
. twoway scatter bweight gestwks
```

The plot suggests that there is a roughly linear association between birth weight and gestation period. To change the symbol which marks the points to a small x, try

```
. twoway scatter bweight gestwks, msymbol(smx)
```

For a list of marker symbols and sizes, try

```
. graph query symbolstyle
. graph query markersizestyle
```

To produce separate plots for boys and girls, try

```
. twoway scatter bweight gestwks, msymbol(smx) by(sexalph)
```

5.4   Overlaying graphs

Plots produced for different subsets of the data, like baby boys and girls, can be overlaid. For example, try

```
. twoway scatter bweight gestwks if sex==1,
      msymbol(smcircle) mcolor(blue)
```

to produce the scatter plot of `bweight` against `gestwks` for boys, and

```
. twoway scatter bweight gestwks if sex==2,
      msymbol(smx) mcolor(red)
```

to produce the scatter plot for girls. To overlay these, try

```
. twoway (scatter bweight gestwks if sex==1,
      msymbol(smcircle)  mcolor(blue))
      (scatter bweight gestwks if sex==2,
      msymbol(smx) mcolor(red))
```

The legend is not helpful, so add the option

```
      , legend(label(1 "Boys") label(2 "Girls"))
```

where the number 1 stands for the first key and 2 for the second. Note that quotes are needed here. Be very careful with the comma as this option applies to the whole graph. The syntax when overlaying two or more twoway plots is

```
. twoway ( ... , options ) ( ... , options) , options
```

where the first `options` refers to the first plot, the second `options` refers to the second plot, and the third to the overall graph.

5.5   Line plots

The dataset `meanbw.dta` is derived form the births dataset, and holds the mean and standard deviation of the birth weights of babies by `sex` and `agegrp` where `agegrp` contains the mid-point of the interval for maternal age to which the mother belongs, using the intervals (20–30), (30–35), (35–40), and (40–45). Load the data and have a look at them with

```
. use meanbw, clear
. list, separator(4)
```

To plot the mean birth weights against age group as a scatter plot, try

```
. twoway scatter meanbw agegrp
```

To label the points using the values of the variable `sex`, try

```
. twoway scatter meanbw agegrp, mlabel(sex)
```

If the variable has value labels these will be used instead of the values. To connect the points on the scatter plot, try

```
. twoway connected meanbw agegrp, sort(sex agegrp) mlabel(sex)
```

With this command each point is connected to the next in the sort order, hence the need to sort by `sex` and `agegrp`. You will see that the connecting has worked well apart from the fact that the last point for males is connected to the first point for females. This can be avoided by using the option `connect(ascending)`, or `connect(L)` for short, which only connects points while the X values are ascending:

```
. twoway connected meanbw agegrp,
         sort(sex agegrp) connect(ascending) mlabel(sex)
```

To show only the lines, try

```
. twoway line meanbw agegrp,
         sort(sex agegrp) connect(ascending)
```

The difference between `line` and `connected` is that `line` does not show marker symbols for the points which are joined by lines, while `connected` does.

It is sometimes useful to add confidence intervals to plots of means. Try

```
. generate lower = meanbw - 1.96 * stdev
. generate upper = meanbw + 1.96 * stdev
```

to create the upper and lower bounds. To plot these against `agegrp` for boys, joining them with vertical capped lines, try

```
. twoway rcap lower upper agegrp if sex==1
```

To show both boys and girls it is better to move the girls up by 1 on the agegrp axis, so the plots can be distinguished. Try

```
. replace agegrp=agegrp+1 if sex==2
. twoway rcap lower upper agegrp
```

To overlay this plot with the connected mean values, try

```
. twoway (rcap lower upper agegrp)
         (connected meanbw agegrp, sort(sex agegrp) connect(L) mlabel(sex))
```

5.6 Cumulative distribution plots

Like box plots, cumulative distribution function (cdf) plots (also called cumulative probability plots) are better than histograms for metric data because they don't require you to choose the number or width of the bins. Stata 8 has no command for producing cumulative distribution plots directly, although a number have been contributed by users. Try

```
. use births, clear
. cdf bweight
```

which uses Stata 7 graphics.[1] Note the staircase effect when `cdf` is used with a variable such as `matage` which takes only a limited number of values:

```
. cdf matage
```

Although Stata 8 options don't usually work with Stata 7 commands, in this case the labelling of the X-axis and the Y-axis can be improved with

```
. cdf matage, xlabel(25 30 35 40 45) ylabel(0 0.5 1)
```

This type of plot is useful when comparing several distributions on the same graph. For example,

```
. cdf bweight, by(hyp)
```

shows two cumulative distributions, one for normal mothers, one for hypertensive mothers.

To produce a cumulative distribution plot of `matage` with Stata 8 graphics it is necessary first to use the command `cumul` to calculate the cumulative relative frequencies and place them in a variable called (for example) `cdf`:

```
. cumul matage, generate(cdf) equal
```

The option `equal` forces observations that share the same value of `matage` to have the same cumulative relative frequency. The cumulative relative frequencies are then plotted against the values of `matage` using a line graph:

```
. twoway line cdf matage, connect(stairstep) sort
```

The `connect` option in this example states that the points are to be connected using a stairstep which is a horizontal line followed by a vertical line. Try `graph query connect` for a full list of connect options.

---

[1]The command `cdf` is a not part of official Stata, but it is included with the files which come with the book (see Chapter 0).

5.7 Adding lines

Vertical lines can be added through any point on the X-axis with `xline( )`. Similarly
horizontal lines can be added with `yline( )`. These options can be repeated to add
several lines. As an example, try

. twoway scatter bweight gestwks, xline(39) yline(2500 3000) yline(3500)

It is often useful, with a scatter plot, to overlay the best fitting straight line. For
example, try

. twoway scatter bweight gestwks

to produce the scatter plot of birth weight versus gestation period, and

. twoway lfit bweight gestwks

to plot the best fitting straight line. To overlay the two graphs, try

. twoway (scatter bweight gestwks) (lfit bweight gestwks)

and to see the confidence interval about the fitted line, try

. twoway (scatter bweight gestwks) (lfitci bweight gestwks)

5.8 Graph titles

Stata allows four possible sorts of title for a graph: title, subtitle, caption, and note.
To see where they go, and their relative sizes, try

. twoway scatter bweight gestwks,
        title(TITLE) subtitle(SUBTITLE) caption(CAPTION) note(NOTE)

When a graph command is combined with `by`, the graph is repeated for each category
of the variable in the `by`. Try

. twoway scatter bweight gestwks,
        title(TITLE) subtitle(SUBTITLE) caption(CAPTION) note(NOTE)
        by(sexalph)

and you will see two graphs with identical titles. When the subtitle is not set that
position is used to identify the graphs using the values of `sexalph`. Try

. twoway scatter bweight gestwks, by(sexalph) title(TITLE)

and you will see that each graph has the title 'TITLE', and the graphs are identified
in the subtitle position using the values of `sexalph`. To make the title refer to the
entire graph it must be inside the `by`, as in

```
. twoway scatter bweight gestwks, by(sexalph, title(TITLE))
```

If you want to set your own overall note this must also be within the `by`, as in

```
. twoway scatter bweight gestwks, by(sexalph, title(TITLE) note(MYNOTE))
```

The automatic subtitles and notes can be removed using `subtitle("")` outside the `by( )` and `note("")` inside the `by( )`.

When the variable in the `by( )` is numeric, the values it takes are used to identify the graphs. For example, try

```
. twoway scatter bweight gestwks, by(sex)
```

To replace the numeric codes with text first label the values of `sex`:

```
. label define sex 1 "Boys" 2 "Girls"
. label values sex sex
. twoway scatter bweight gestwks, by(sex)
```

Check that the graphs are now identified as Boys and Girls. For convenience a number of generally applicable options like `title` are gathered together in Table 5.2.

5.9   Titles and labels for axes

Axis titles are the words appearing alongside an axis and axis labels are the numbers which appear by the tick marks. The default title for an axis is the variable label, or failing this, the name of the variable plotted on that axis. After

```
. describe
. twoway scatter bweight gestwks
```

you will see that the X-axis has the title 'gestation period', which is the label for the variable `gestwks`, while the Y-axis has the title 'birth weight'. To change the size of the default title for the X-axis, try

```
. twoway scatter bweight gestwks, xtitle(, size(large))
```

and to change the default title as well, try

```
. twoway scatter bweight gestwks,
        xtitle(Gestation period in weeks, size(large))
```

Quotes are not required for titles.

The default labels for an axis are the best round numbers that cover the range of values taken by the variable. To change the default for the Y-axis, try

```
. twoway scatter bweight gestwks, ylabel(1000(500)5000)
```

Group	Option
Graph titles	`title(text, size( ))`
	`subtitle(text, size( ))`
	`caption(text, size( ))`
	`note(text, size( ))`
with by	place the above options inside the by( )
Axes	`xtitle(text, size( ))`
	`ytitle(text, size( ))`
	`xlabel(numlist, labsize( ) angle( ))`
	`ylabel(numlist, labsize( ) angle( ))`
	`xscale(range(numlist) log)`
	`yscale(range(numlist) log)`
Added lines	`xline(#, lpattern( ) lcolor( ))`
	`yline(#, lpattern( ) lcolor( ))`
Marker symbols	`msymbol( ) msize( ) mcolor( ) mlabel( )`
Connect style	`connect( )`
Legends	`legend(label(# "text") label(# "text") ...)`
	`legend(order(# "text" # "text") ...)`

Table 5.2: Some general options for graphs

The expression 1000(500)5000 is short for 1000,1500,2000, ... , 5000, and is an example of a number list. You can change the size of the labels and the angle with

```
. twoway scatter bweight gestwks,
        ylabel(1000(500)5000, labsize(large) angle(horizontal))
```

5.10   Naming, saving, and combining graphs

Graphs can be saved either in memory or on disk. Repeat the scatter plot of birth weight against gestation period with

```
. twoway scatter bweight gestwks
```

This graph can be recalled at any time by clicking on the *Bring Graph Window to Front* icon, but when you create a new one it will over-write the first one. To keep the graph more permanently you need to name it. To name it `mygraph`, try

```
. twoway scatter bweight gestwks, name(mygraph)
```

The graph can now be displayed at any time with

```
. graph display mygraph
```

To replace a graph with a new one bearing the same name, try

. twoway scatter bweight matage, name(mygraph, replace)

Graphs can also be saved (with a name) to disk. For example

. graph save mygraph

will save the current graph as mygraph.gph where gph is the default extension for Stata graphs. If a file with that name already exists, add the option replace as in

. graph save mygraph, replace

Any saved graph can be retrieved at a later time with the command graph use. For example try

. graph use mygraph

You can combine different graphs with the command graph combine either from memory or from disk. For example

. twoway scatter bweight matage, name(bw_age)
. twoway scatter bweight gestwks, name(bw_gest)
. graph combine bw_age bw_gest

will combine the two named graphs into a single graph showing the two scatter plots side by side. Alternatively, try

. twoway scatter bweight matage, saving(bw_age)
. twoway scatter bweight gestwks, saving(bw_gest)
. graph combine bw_age.gph bw_gest.gph

which will combine the two graphs from disk. Note that the extension .gph is required when using graph combine with graphs saved on disk. Any number of graphs can be combined.

5.11   Printing and exporting graphs

Whenever a graph is produced it can be printed using the *File* tab and then *Print Graph*, or you can click on the *Print* icon. Graphs can also be copied directly onto the clip-board using the *Edit* tab and then *Copy Graph*. They can then be pasted into documents created by most wordprocessors.

To export a current graph as an .eps file, called, for example, mygraph.eps use file name extensions!.eps

. graph export mygraph, as(eps)

You can use other possibilities for the as( ) option, like pdf, which will save the graph as a pdf file. For a full list of the export possibilities, try

. help graph_export

5.12   Schemes

Schemes govern the default appearance of graphs. The scheme which is set when Stata 8 is first installed is called `s2color`. For a list of available schemes, try

`. help schemes`

To see what the economist scheme looks like, try

`. twoway scatter bweight gestwks, scheme(economist)`

You can change the scheme permanently to economist with `set scheme economist`.

5.13   Help for graphics

Help may be required on many different aspects of a `graph` command, and it is not always clear how to get it. When in doubt, start with `help graph_intro`. A list of some other useful sources of help is given below.

Command	Help for
`graph query symbolstyle`	List of main plotting symbols
`graph query markersizestyle`	List of sizes for plotting symbols
`graph query colorstyle`	List of colors for plotting symbols
`graph query linepatternstyle`	List of patterns for lines
`help title_options`	Graph titles
`help axis_options`	Axes
`help added_line_options`	Added lines
`help marker_options`	Marker symbols
`help connect_options`	Connecting points
`help legend_option`	Legends

Exercises

1. Load the births data and obtain box plots of `gestwks` for each category of `lowbw`. Add a title and improve the labelling of the Y-axis.

2. Plot a histogram for the metric variable `gestwks`. Label the X-axis from 20 to 45 using intervals of 5. Start the first bin at 20, and use 5 as the bin width.

3. Tabulate the categorical variable `sex` and obtain the corresponding histogram. Label the vertical axis from 0 to 1 with intervals 0.2 and improve the X-axis.

4. Obtain a scatter plot of `gestwks` vs `matage` together with the best fitting line.

Chapter 6

Graph dialog boxes

There is a lot more to Stata 8 graphs than we have shown in the preceding chapter, and most of this is best explored using the graph dialog boxes. The first item under the *Graphics* tab is called *Easy graphs* and this leads to a list of possibilities ranging from *Scatter plot* to *Function graph* all of which call up easy dialog boxes. The rest of the items under the *Graphics* tab call up the full dialog boxes. The easy dialog boxes offer fewer options than the full dialog boxes, and are simpler to use. Full dialog boxes can be called up directly with (for example)

```
. db histogram
```

and the corresponding easy dialog box can be called up with

```
. db ehistogram
```

where 'e' stands for 'easy'. In this chapter we shall concentrate on the full dialog boxes, and describe these for some common graphs. It is not our intention to provide a complete survey of all the graph dialog boxes, only to introduce a few of them so that the general style becomes familiar.

6.1 Histograms

Start with

```
. use births, clear
. db histogram
```

and press R to remove any previous selections.

The main tab

To make a histogram of birth weight, enter `bweight` in the *Variable* box, check *Width of bins* and enter 500, check *Lower limit of first bin* and enter 0, and check *Percent*. Press *Submit* to produce

```
. histogram bweight, width(500) start(0) percent
```

If you check *Discrete data*, for the histogram of a categorical variable, the number of bins will be the same as the number of categories. Experiment with the rest of the controls in this tab which refer to the color used for filling and outlining the bins (or bars), and return to the defaults when you have finished, but leave the selections for *start*, *width*, and *Percent*.

The title and captions tabs

Select the *Title* tab and enter `TITLE` for the title and `SUBTITLE` for the subtitle. Select the *Caption* tab and enter `CAPTION` as the caption and `NOTE` as the note. Press *Submit* to see where these appear. Experiment with different sizes, colors, and positions for each of them, and then remove all titles and return to the defaults for this tab when you have finished.

The overall tab

The controls in this tab are concerned with the overall appearance of the graph and allow you to change the scheme, and to name the graph so that it is stored in memory. To see the difference between the different regions (inner and outer for graph, inner and outer for plot) select a different fill color for each of the regions, and see what happens. You can also experiment with different colors for the outlines. Return to the default values for this tab when you have finished experimenting.

The by tab

Enter `sexalph` in the *Variables* box and press *Submit* to produce

```
. histogram bweight, width(500) start(0) percent by(sexalph)
```

You will get one graph for each category of `sexalph`. Checking *Graph total* includes a total graph and checking *Graph missing* includes a graph based on observations for which `sexalph` is missing. *X-Rescale* and *Y-Rescale* refer to whether or not the X and Y axes are rescaled for each graph defined in the *By* tab (the default is not), and *Scale text* allows the text size to be scaled by a factor such as 1.5.

   When you enter `sexalph` in the *Variables* box of the *By* tab the controls in the *Overall*, *Title* and *Caption* tabs refer to the individual graphs for each category of `sexalph`. To control the overall appearance of the graph you use the *Graph region* and *Title/Subtitle/Caption/Note* controls in the *By* tab. Unless you fill in something for the *Note* it is used for the phrase 'Graphs by sexalph ', or whatever is the name of the variable defined in the by tab. If you don't want this, just fill in "" for the note. The final point to note is that when you use the *By* tab the *Subtitle* of the *Titles* tab, if set, over-rides the default identification of the individual graphs. A good experiment is to fill in TITLE, SUBTITLE, CAPTION, and NOTE in the appropriate boxes of the *Title* and *Caption* tabs, and AAA, BBB, CCC, and DDD in the *Title/Subtitle/Caption/Note* controls of the *By* tab. Pressing *Submit* will show where they all go. Now remove SUBTITLE and DDD, and you will see the default use for these two when nothing is set.

The x-axis and y-axis tabs

These two tabs contain identical controls. Select the *X-axis* tab, enter `Birth weight in grams` for the *Title*, and check *Box* if you want the title to be boxed. The scale options allow you to choose numbers which the scale must cover. Under *Ticks/Lines* enter the number list `0(500)5000` in the *Rule* box and press *Submit* to produce

```
. histogram bweight, width(500) start(0) percent
  xtitle(Birth weight in grams) xlabel(0(500)5000)
```

The *Alt labels* control asks for two rows of alternate labels. The *Custom* box is useful for alphabetic labels such as months. The *Grid* box controls the grid which is superimposed on the graph (the default is a horizontal grid through the major ticks). Be warned that when you check *Range* under *Scale options* a decimal point, almost invisible to the naked eye, appears in each of the two boxes alongside.

The if/in, weights, normal density, KDE, and bar label tabs

These are straightforward. The *Weights* tab shows which sort of weights are allowed (only frequency weights with histogram) and KDE refers to a kernel density estimation, a method for producing a smooth density which fits the histogram locally.

The legend tab

This tab will be described under box plots.

6.2 Box plots

Start by closing the dialog box for histograms and calling up the one for for box plots. Press R to remove any previous selections.

The main tab

Enter `bweight` in the *Variables* box and press submit to produce

`. graph box bweight, medtype(line)`

You can choose between *Line, Custom line* and a *Marker* to indicate the median. Checking one or other of these last two brings up some further options. Select *Custom line* with *dash* as the *Pattern*, and press *Submit* to produce

`. graph box bweight, medtype(cline) medline(lpattern(dash))`

The over tab

Select the *Over* tab, enter `hyp` in the *Over 1 Variable* box and `sexalph` in the *Over 2 Variable* box. Now enter `1 "Normal" 2 "Hypertensive"` in the *Relabel* box for `hyp` and `1 "Boys" 2 "Girls"` in the *Relabel* box for `sexalph`. Press *Submit* to see four box plots corresponding to the four combinations of `hyp` and `sexalph`. The box plots can be sorted in various ways; you can change the order to Girls then Boys, for example, by checking *Descending*. The *Categorical axis* is the one defined by the `over` variables, and can be labelled in different colors and sizes.

The boxes tab

This tab can be used to change the fill color for the boxes, but note that when only one Y variable is specified only one fill color can be selected. The others make no difference. You can also specify the outer gap and the box gap. Experiment with an *Outer gap* of 50 and a *Box gap* of 0, then reverse these to see the effect.

   The box plots of birth weight over hypertension and sex could be improved by color coding boys and girls. Because it is not possible to use different colors for the boxes when using only one Y variable we must use another feature which allows multiple Y variables to be specified. Start by generating two new variables, one for the birth weight of boys and one for that of girls, with

`. separate bweight, by(sex)`

to produce `bweight1` which contains the birth weight for boys, and `bweight2` which contains the birth weight for girls. Then select the *Main* tab to enter `bweight1` and `bweight2` in the *Variables* box instead of `bweight`. Leave the *Over* tab as it is, and return to the *Boxes* tab. Select *Red* for the *Fill color* and the *Line color* for *Box 1* and *Green* for *Box 2*. Then click on the *Outsides* tab, and select *Red* for the *Color* of *Marker 1* and *Green* for *Marker 2*. Press *Submit* to see the color coded boxes.

The legend tab

Stata has done its best to produce a legend based on the variable labels for `bweight1` and `bweight2`, but this can be improved. Select the *Legend* tab and enter `1 "Boys"`

2 "Girls" in the *Label* box (on the right hand side). Press *Submit* and you will see a better legend. The *Title/Subtitle/Caption/Note* group of controls refers to titles within the box containing the legend. Similarly the *Legend controls* refer to how the keys within the legend box are arranged, and the *Region options* allow a choice of fill colors for the legend box.

The outsides tab

These controls refer to the points outside the boxes. As with the *Boxes* tab, when there is only one Y variable only one set of these controls does anything. When there are two, two sets can be used, and so on. Each set allows you to change the marker symbol used for the outside points, and its size and color.

The rest of the tabs

These all behave like their equivalents in the histogram dialog box, except for the *Misc.* tab which is a catch-all for some minor points.

6.3 Bar charts

Bar charts in Stata need a variable for each bar. To make a bar chart showing the mean birth weight for boys and girls, close the dialog box for box plots and bring up the one for bar charts with `db bar`. Enter `bweight1` in the first *Variable* box, after the = sign, and `bweight2` underneath in the next *Variable* box. Press submit to produce

```
. graph bar (mean) bweight1 bweight2
```

Instead of *Mean* you could have selected *Median* or any other possibility in the pull down menu under *Mean*. As it stands it is not a very interesting bar chart, but we could make it more interesting by selecting the *Over* tab and entering `hyp` for *Over 1*. Press *Submit* to see the result. The labelling could be improved by entering 1 "Normal" 2 "Hypertensive" in the *Relabel* box of the *Over* tab. Similarly the legend could be improved by entering 1 "Boys" 2 "Girls" in the *Label options* group of controls of the *Legend* tab.

6.4 Twoway graphs

Start with

```
. db twoway
```

and you will be offered a choice between *Twoway graphs (scatterplot, line, etc.)* and *Overlaid twoway graphs*. Choose the former.

The plot 1 tab

Enter `gestwks` as the *X-variable* and `bweight` as the *Y-variable*, and press *Submit* to produce

```
. twoway (scatter bweight gestwks)
```

The parentheses are not necessary, but do no harm. The pull down menu under *Type* shows a long list of possible types of twoway plot. The *Markers* group refers to the symbols used for the points in the plot, and clicking the *Markers label* group allows you to enter the name of the variable whose values will be used to label the points. The *if* box allows the plot to be restricted (but there is no *in*), and the *Additional graph options* box is a catch-all which is unlikely to be of any use to a beginner.

6.5   Overlaid twoway graphs

Close the dialog box and start again with

```
. db twoway
```

and select *Overlaid twoway graphs*.

The plot tabs

The *Plot 1* tab here is the same as in the twoway scatter dialog box, and refers to the first of the plots to be overlaid. The *Plot 2* tab is initially almost completely blank, but selecting a plot *Type* produces a page like *Plot 1* which refers to the second plot, and so on up to *Plot 4*. In the *Plot 1* tab select `scatter` as *Type*, and then enter `gestwks` in the *X* box and `bweight` in the *Y* box. Restrict the plot to boys by entering `sex==1` in the *if* box. In the *Plot 2* tab select `scatter` as *Type*, and then enter `gestwks` in the *X* box and `bweight` in the *Y* box. Restrict the plot to girls by entering `sex==2` in the *if* box. Press *Submit* to produce

```
. twoway (scatter bweight gestwks if sex==1)
         (scatter bweight gestwks if sex==2)
```

As another example, remove the `sex==1` and `sex==2` from the *Plot 1* and *Plot 2* tabs, and select *lfit* in the pull down menu for *Type* in the *Plot 2* tab. Press *Submit* to produce the command

```
. twoway (scatter bweight gestwks) (lfit bweight gestwks)
```

The r-axis tab

This is used when two Y variables are specified, one for *Plot 1* and one for *Plot 2*. For example, enter `matage` for the *Y-variable* in *Plot 2* and select *scatter* as *Type*.

Check the *Plot on right axis* box. The *R-Axis* tab now controls the right hand axis where the Y-values of `matage` are plotted. Select this tab and enter `Maternal age in years` in the *Title* box. Press submit to produce

```
. twoway (scatter bweight gestwks)
        (scatter matage gestwks, yaxis(2)),
        ytitle(Maternal age in weeks, axis(2))
```

Exercises

There are no exercises for this chapter.

Chapter 7

More basic tools

In this chapter you will learn how to use the return list, how to generate random numbers, how to group the values of a metric variable, how to compare two means or proportions, how to use weights, how to repeat commands for different groups of observations using the `by` and `bysort` commands, and how to repeat commands using `foreach`.

7.1 The return list

After each Stata command some of the results are stored in memory so that they can be referred to. For example, try

```
. use births, clear
. summarize bweight
. return list
```

The mean is returned in `r(mean)`, the standard deviation in `r(sd)` the sum in `r(sum)`, and the number of observations in `r(N)`. To refer to the values, try

```
. display r(sum)
. display r(N)
```

For a more extensive return list try

```
. summarize bweight, detail
. return list
```

For a very brief one, try

```
. tabulate hyp sex
. return list
```

You will see that `r(r)` and `r(c)` return the number of rows and columns of the table.

Returns are useful when you need the results from one command to be fed into a new command. For example, to create a new variable `dbw` that holds the difference between each baby's birth weight and the overall birth weight mean we could use

```
. summarize bweight
. generate dbw =  bweight - r(mean)
```

7.2   Generating variables using functions

Stata functions can be used with `generate` and take the general form

$$generate\ new\_var = fname(old\_var)$$

where *new_var* is the name of the new variable you wish to generate, *old_var* is a variable already in your dataset, and *fname* is the name of the function. For example,

```
. generate sqrbw=sqrt(bweight)
```

It is important not to leave a space between the function name and the first parenthesis, otherwise Stata thinks *fname* is a variable name, not a function.

If you ask for help on functions by selecting *Stata command* from the *Help* menu, you will see that they are arranged in 8 groups, starting with mathematical functions. Some of the basic functions like `ln( )` have already been introduced, but another useful one is `uniform( )`, which produces random numbers between 0 and 1. Note that although `uniform( )` is a function it does not require an argument but it still needs the parentheses ( ). Try

```
. generate u=uniform( )
. list u
. cdf u
```

The cumulative distribution of u shows that its values are uniformly distributed. The function `norm( )` returns the standard normal cumulative probability, and its inverse is `invnorm( )`. To convert uniform random numbers between 0 and 1 into numbers with a standardized normal distribution try

```
. generate n=invnorm(u)
. cdf n
```

Functions can also be used as part of expressions used with `generate`, as in

```
. generate x=sqrt(10)*uniform( )
```

Another class of functions can be used with the command `egen` which stands for extensions to generate. These take the general form *egen new_var = fname(old_var)*

However, it is not possible to use **egen** functions as part of expressions. Some examples of **egen** functions are shown in the next section. Ask for help on **egen** to see the list of available functions.

7.3   Grouping the values of a variable

When a variable has many values, like **matage**, it is often useful to group the values and to create a new variable which codes the groups. For example we might cut the values taken by **matage** into the groups 20–29, 30–34, 35–39, 40–44, and to create a new variable called **agegrp** coded 20 for subjects in the first age group, 30 for subjects in the second age group, 35 for subjects in the third age group, and so on, using the lower end of the age group as the code. The best way of doing this is to use **egen** with the function **cut**. Try

```
. egen agegrp=cut(matage), at(20,30,35,40,45)
. browse
```

You will see that the new variable **agegrp** takes the value 20 when **matage** is in the first age group, 30 when **matage** is in the second age group, and so on. The last group, consisting of values from 40 up to but not including 45, is coded 40.

The function **cut** offers a number of useful options. If you prefer to code the groups using the integer codes 0, 1, 2, ... instead of 20, 30, 35, ... try

```
. drop agegrp
. egen agegrp=cut(matage), at(20,30,35,40,45) icodes
. tab agegrp
```

It is necessary to drop **agegrp** first because **egen**, like **generate**, creates a new variable. To save some typing you could replace (20,30,35,40,45) with (20,30(5)45), both of which are examples of what Stata calls *number lists*.Try **help numlist** for a full list of which abbreviations are allowed in a number list.

You can also label the values of the categorical variable **agegrp** using

```
. drop agegrp
. egen agegrp=cut(matage), at(20,30(5)45) icodes label
. tabulate agegrp
```

It is important to realize that observations which are not inside the range specified in the **at()** part of the command result in missing values for the new categorical variable. For example, try

```
. drop agegrp
. egen agegrp=cut(matage), at(20,30,35) icodes label
. tabulate agegrp
. tabulate agegrp, missing
```

Only observations from 20 up to, but not including 35, are included. For the rest, `agegrp` is coded missing. If you don't want to choose the cut-points for the categories of `matage`, try

```
. drop agegrp
. egen agegrp=cut(matage), group(5)
. tabulate agegrp
```

which will produce 5 roughly equal–frequency groups coded 0, 1, 2, 3, 4. The option `label` will show the lower ends of the intervals in terms of the original units of `matage`. Try this option now (remember to drop `agegrp` first). To create a new variable, `agegrpm`, holding the mean value of `matage` corresponding to each category of `agegrp`, the `egen` function `mean` can be used. Try

```
. egen agegrpm=mean(matage), by(agegrp)
. tabulate agegrpm
```

For a quick reference to the functions available with `egen`, try

```
. db egen
```

7.4   Comparing two means or two proportions

Testing the difference between two means using the t-test is a very basic statistical requirement, and will be illustrated by comparing the mean birth weight of the babies of hypertensive and normal mothers. The command

```
. ttest bweight, by(hyp)
```

shows an unusual amount of output for such a simple operation. The most important line in the table is the last which refers to the difference in the two means. The null hypothesis which is being tested is that the true values of the two means are equal, and the value of t and the corresponding two-sided p-value are shown at the bottom, mid-screen. The t is very large (5.455) and the p-value is small, so the difference is highly significant.

The proportions of low birth weight babies born to hypertensive and normal mothers are obtained with

```
. tabulate lowbw hyp, col
```

The proportions are 9% for normal mothers and 28% for hypertensive mothers. To test the significance of this difference use the option `chi2`:

```
. tabulate lowbw hyp, col chi2
```

The chi-squared statistic is large, and the p-value is small, so the difference is highly significant. Another way of making comparisons between means and proportions which generalizes to several groups is described in Chapter 11.

A few Stata commands are also available in an immediate form, so they can be used in calculator mode, with numbers instead of variables. For example, the test described above could also be carried out with

```
. tabi 388 52 \40 20, col chi2
```

See help for `immed` for a list of immediate commands in official Stata.

7.5   Weights

In some datasets each record has an associated weight. This might be a frequency weight which indicates that the record occurred with a given frequency, or an importance weight which states that when calculating things like means the record should be weighted by its importance. Frequency weights must be integers, but importance weights can be any positive numbers. There are two other kinds of weight in Stata which do essentially the same thing as importance weights, namely probability weights and analytical weights. See the *Stata User's Guide* (14.1.6) for a discussion of the different sorts of weight.

Most Stata commands can be used with weights. For example, suppose the births data consisted of just the variables `lowbw` and `hyp`. There are only four possibilities for the records, namely

```
lowbw hyp
    0    0
    1    0
    0    1
    1    1
```

Rather than enter 500 records of this kind, it would be better to enter the four records

```
lowbw hyp     N
    0    0   388
    1    0    40
    0    1    52
    1    1    20
```

where the last variable, N, gives the frequency with which each combination occurs in the data. Such a dataset is in the file `births_freq.dta`, so try

```
. use births_freq, clear
. tabulate lowbw hyp
```

This ignores the frequencies, and the table is based on just the 4 records. To take the frequencies into account, try

```
. tabulate lowbw hyp [fw=N]
```

where `fw` stands for frequency weights. This shows that `tabulate` now understands that there are 500 observations in the data, not 4. Alternatively, the data can be expanded to individual records, as follows:

```
. expand N
. drop N
. describe
```

There are now 500 records. The command `expand N` works by replacing each record with n copies where n is the value of the variable N for that record. For example, for the first record N takes the value 388, so the first record is replaced by 388 copies of itself.

7.6   Repeating commands for different sub-groups

Stata has a powerful facility for processing records by groups. For a straightforward example, load the births data and summarize the variable `bweight` in detail, for each hypertension group, as follows:

```
. use births, clear
. sort hyp
. by hyp: summarize bweight, detail
```

Of course this could have been done just as easily with

```
. summarize bweight if hyp==0, detail
. summarize bweight if hyp==1, detail
```

but spelling out the different values of the variable which is defining the groups would get tedious with more than 3 or 4 values, and the `by` command can be used to automate this. The groups can also be defined by combinations of variables, as in

```
. sort hyp sex
. by hyp sex: summarize bweight, detail
```

To include the `sort` with the `by` try

```
. bysort hyp sex: summarize bweight, detail
```

7.7   Repeating commands for different variables

Some Stata commands allow you to specify a list of variables, and will repeat the
command for each variable. For example

```
. summarize matage bweight gestwks
```

summarizes each of the variables in the list. You could do the same thing with

```
. foreach var of varlist matage bweight gestwks {
. summarize 'var'
. }
```

Here `foreach` first defines the list of variables that are to be used in sequence, using
the keyword `varlist`. Then it lists, within curly brackets, all the commands that need
to be repeated for each variable in the variable list. The expression `'var'` within the
curly brackets refers to each of these variables, in turn. Any name can be used. For
example

```
. foreach X of varlist matage bweight gestwks {
. summarize 'X'
. }
```

would do as well.

The single quotes which surround `var` and `X` are important – the left hand single
quote is different from the right hand one. On most keyboards you will find them on
the top left-hand corner (below the Esc key) and near the Enter key of the keyboard,
respectively (the full meaning of these quotes will be explained in Chapter 19). If you
are using a non-English keyboard you may not find these keys. In this case it is best
to allocate two of the function keys, as follows:

```
. macro define F4=char(96)
. macro define F5=char(39)
```

Now pressing F4 will produce the left-hand quote and F5 will produce the right-hand
quote.

You can repeat several commands by enclosing them in the brackets { }, but each
command must be typed on a new line, as in

```
. foreach var of varlist matage bweight gestwks {
. summarize 'var'
. count if 'var' >= .
. }
```

Stata prints a new line number for each new line, a process which normally stops with
the last }, but if you make errors, and it won't stop, type `end`.

You can also use `'var'` within the text displayed by `display` as in

```
. foreach var of varlist matage bweight gestwks {
. summarize `var'
. display "Mean of variable `var' is  " r(mean)
. }
```

The keyword `varlist` is used to indicate a list of existing variables. If you want to create new variables the keyword is `newlist`. For example,

```
. foreach var of newlist tom dick harry {
. generate `var' = bweight
. }
. list tom dick harry in 1/10
```

will generate 3 copies of `bweight` called `tom`, `dick`, and `harry`.

The command `foreach` also works with lists of numbers when the keyword `numlist` is used. Try

```
. foreach num of numlist 1,2,3 {
. generate bw`num'=bweight
. }
. list bw* in 1/10
```

which creates three new variables `bw1`-`bw3`, each a copy of `bweight`. Make sure there is no space between `bw` and `num`. Their names are made of the common root `bw` followed by one of the numbers in `numlist`.

These examples refer to different kinds of list – `varlist` refers to existing variable names, `newlist` refers to new variable names, and `numlist` refers to a Stata number list. Using the syntax `of varlist`, or `of newlist`, or `of numlist` enables Stata to carry out some rudimentary checks, for example that the list of names in a `varlist` are all existing variable names. A looser syntax can be used with any kind of list. For example,

```
. foreach var in matage bweight gestwks {
. summarize `var'
. }
```

will have the same effect as

```
. foreach var of varlist matage bweight gestwks {
. summarize `var'
. }
```

without the checks. This more general kind of list is useful with file names. Try

```
. foreach file in births dates {
. use `file' , clear
. describe
. }
```

Each of the data files is loaded and described, in turn. Note that the keyword `of` was used for lists of variables and numbers, but `in` is used for a general list such as one of file names. See Cox (2001)[4] for a good review of how to repeat commands in Stata.

Exercises

1. Load the births data and create a new variable `gest4` which cuts `gestwks` at 20, 35, 37, 39, 45 weeks using `egen`.

2. Obtain a frequency table for the categorical variable `gest4` using `tabulate`.

3. Obtain a table of mean birth weight by categories of `gest4`, using `table`.

4. Create a new variable `diffma` equal to the difference between `matage` and the overall mean of `matage` using `r(mean)`.

5. Create a new variable equal to the square root of `bweight`.

6. Use `bysort` to summarize `bweight` for each value of `sex`.

7. Create the categorical variable `mage4` from `matage` using `egen` with the function `cut` and the options `group(4)` and `label`. Tabulate `mage4`: what are the quartiles used to group `matage`?

8. Test whether the birth weight means of boys and girls are significantly different using the command `ttest`.

9. Tabulate each of the variables `hyp`, `sex` and `preterm` using the command `foreach`.

10. Summarize, take logs, and then summarize the log-transformed variable, for each variable in the list `matage gestwks bweight`, using `foreach`. You can either replace each variable by its log transform (bad practice) or create a new variable, say `log'var'`, to hold the log transformed variable before summarizing. Don't forget to drop `log'var'` after summarizing.

Chapter 8

Data management

In this chapter you will learn how to clean data; how to append one data file to another; how to merge data files, and how to update one file with information from another.

8.1 Cleaning data

Cleaning data means eliminating any errors which occurred while the data were being collected or computerized. It involves making checks on the values which the variables take, and is best done with a do file. We shall give some examples of common checks, using the births data. The commands

```
. use births, clear
. summarize
```

can reveal some discrepancies, but more specific checks may be needed. To check that sex takes only the values 1, 2, or missing, we can count and identify any observations where this is not true with

```
. count if sex!=1 & sex!=2 & sex<.
. list id sex if sex!=1 & sex!=2 & sex<.
```

We have excluded missing values with sex<. rather than sex!=. to make sure that all missing codes are excluded. The result of the first command should be 0 if sex is coded correctly; the result of the second should list the id and the value for sex for any subjects who have been incorrectly coded. Let us recode sex as 3 for subject 10, and run the checks again:

```
. replace sex=3 if id==10
. count if   sex!=1 & sex!=2 & sex<.
. list id sex if   sex!=1 & sex!=2 & sex<.
```

Once an error has been detected it can be corrected using the Data Editor or the command `replace`.

To check that `gestwks` lies in the range 20–44 if it is not missing, count the observations which are less than 20 or greater than 44 and not missing, as follows:

```
. count if gestwks<20 | (gestwks>44 & gestwks<.)
```

It is always important to bear in mind the possibility that a value may be missing when carrying out checks. If it is missing remember that the missing value codes for numeric data are treated as very large numbers, so in this example a missing value for `gestwks` would be greater than 44.

8.2   String variables

String variables are more difficult to deal with than numeric variables because there are so many more possibilities with strings than with numbers. Fortunately there are good functions in Stata which help with managing string variables. To illustrate some of these, load and list the data in the file `string.dta` with

```
. use string, clear
. list
```

The file contains the names of three subjects in the variable `name`. To sort the file on `name` try

```
. sort name
```

Sorting on a string variable sorts on the first character, then the second, and so on. To create a new string variable containing the name in upper case, try

```
. generate name1=upper(name)
. list
```

We shall now separate the first name from the last and put them together in the opposite order. Start by finding where the blank occurs with

```
. generate b=index(name, " ")
```

The variable `b` contains the position where a blank character first occurs in the variable `name`. Now we can use the `substr` function to separate the first name (from 1 to $b-1$) from the last name (from $b+1$ to the end). Try

```
. generate first=substr(name,1,b-1)
. generate last =substr(name,b+1,.)
. list
```

where `substr(name,b+1,.)` extracts the string from $b + 1$ to the end, wherever that might be. Finally, to put them together in the opposite order, separated by a comma, try

```
. generate name2=last + ", " + first
. list
```

For some commands, where a string variable is not allowed, it is useful to create a numeric variable which takes the value 1 for the first combination of string characters, 2 for the second, and so on. Identical strings are coded with the same number. The command to do this for the variable `name` is

```
. encode name, gen(namecode)
. list name namecode
```

where `namecode` is the numeric variable which contains numerical codes in place of the strings. At first sight it seems that nothing has changed because the values of `namecode` have been labelled with the values of `name`. Try

```
. list name namecode, nolab
```

to convince yourself that `namecode` is numerical, and note that the codes have been allocated in the alphabetic order of the names.

Finally, it is worth mentioning that any number such as 24 can be coded numerically, or as the string composed of the character 2 and the character 4. It sometimes happens that a variable contains numbers as strings, which can be most confusing. When the variable is listed the values look like numbers, but they are not. However, they can easily be converted to numbers using the function `real()` or the command `destring`. For more information about string functions try

```
. help strfun
```

or look up *strings* in the *Stata User's Guide* (16.3.5).

8.3   Appending to add more subjects

Data are often collected separately on subgroups of subjects and stored in different files. The separate files are then appended to each other to form a single file. To illustrate the `append` command we shall start with the file `agesex.dta` which contains

```
    id        age        sex
   100         47          m
   101          .          f
   102         67          m
```

and append the file `newsubj.dta` which contains

```
   id        age        sex
  103         22          m
```

with the following commands

```
. use agesex, clear
. append using newsubj
. browse
```

8.4   Merging to add more variables

Another way of collecting data is to store different kinds of information in different files, and then to merge the files. To make sure that the information is merged
correctly we need a variable which is common to both files and which uniquely identifies the records. As an example we shall merge the file `agesex.dta` with the file
`employ.dta` which contains information on a new variable `employed`:

```
   id    employed
  100         yes
  101         yes
  102          no
```

The common variable which identifies the subjects is `id`. To merge the files, matching
on `id`, both files must be in Stata format (`.dta`) and both must be sorted on `id`, so
first make sure that this is true

```
. describe using agesex
. describe using employ
```

Now try

```
. use agesex, clear
. merge id using employ
. sort id
. list
```

The file in memory before the merge (`agesex.dta`) is called the *master* file while
the file on disk with which it is to be merged (`employ.dta`) is called the *using* file.
The final `sort id` is only there for presentation purposes, because after a merge the
observations are often left in a different order from the order before the merge.

The `list` command shows a new variable called `_merge`. This is created by Stata
whenever the command `merge` is used. It takes the values:

1 when the observation is only from the master file
2 when the observation is only from the using file

3 when the observation is from both.

In this case the value is 3 for all subjects. You should always tabulate _merge after merging to see how many records of each type you have, and to make sure you understand the reason for any 1's and 2's. To keep only those observations made of data from both files, use

```
. keep if _merge==3
```

After convincing yourself that the merge has worked and that you do not need the variable _merge any further you can drop it. If you are merging an additional file you must first drop _merge: otherwise an error message will appear when the new merge tries to create it.

8.5 Merging to update variables

Another use of merge is to update the information on some of the variables in a dataset. For example, the age for subject 101 is missing in agesex.dta. Suppose we wished to update this missing value with the correct age, which is in the file age101.dta containing

```
 id        age
101         32
```

A single correction of this kind could be done more easily with the command replace, but merge would be useful when there were many values to update. Both files are already sorted on id, so try

```
. use agesex, clear
. merge id using age101
. sort id
. list
```

You will see that nothing has changed! Stata carefully guards the master file against change unless specifically authorized with the option update. Now try

```
. use agesex, clear
. merge id using age101, update
. sort id
. list
```

and you will see that the missing value for age has been replaced with its updated value. When update is used the variable _merge takes values 1–5:

1 for an observation from the master file only
2 for an observation from the using file only

3 for an observation from both files, master agrees with using
4 for an observation from both files, missing in master updated
5 for an observation from both files, master disagrees with using file.

In the last of these cases the master would not be updated. Only when the master
value is missing it is updated. If you want to update the master in spite of the
disagreement, use the options `update` and `replace` together.

8.6   Unmatched merges

When no matching variable is specified the `merge` command simply merges the two
files record by record. For example, merging

```
    id         age        sex
   100          47          m
   101           .          f
   102          67          m
```

with

```
employed
     yes
     yes
      no
```

with the commands

```
. use agesex, clear
. merge using employ
```

would give

```
    id         age        sex  employed
   100          47          m       yes
   101           .          f       yes
   102          67          m        no
```

Unmatched merges are not used very often, but it is important to know about them
because it is easy to forget to include the matching variable in the command when
you intend to do a matched merge. What you then get is an unmatched merge, and
it may take a moment before you realize what has happened.

Exercises

1. The file `births_bad.dta` contains the same data as `births.dta`, but with some errors. Can you find them?

2. The file `music.dta` contains the musical skills for 10 subjects. List the `id` for all subject who play the piano. List the `id` for all subjects who play the flute. List the `id` for all subjects who do not play the piano.

3. Data on smoking habits for the mother of the babies in the births dataset are contained in the file `births_smok.dta`. No data are available for 4 women, so the file has 496 records. Start by loading and describing the file `births_smok`.dta. How is the file sorted?

4. How many ever-smokers are there? What is the mean age at which ever-smokers started smoking?

5. Merge the file `births.dta` with `births_smok.dta` matching on `id`.

6. Tabulate the variable `_merge` to see how many women in the study did not have smoking information.

7. List the identifiers for all mothers without smoking information.

Chapter 9

Data management for repeated measurements

Repeated measurement data are also known as panel data, and as longitudinal data. In this chapter you will learn how to use long coding for repeated measurements; how to graph repeated measurements; how to collapse to group level; and how to work at group level without collapsing.

9.1 Wide and long coding

Consider the situation where repeated measurements of one or more variables are made on a subject at different time points during the study. If a variable X is measured on 3 occasions the results could be placed in a single record as

id	X_1	X_2	X_3
1	x	x	x

or in 3 separate records as

id	X	visit
1	x	1
1	x	2
1	x	3

The first of these, where there is a new variable for the measurement at each time point, is called wide coding. The second, where there is one variable for the measurement and one for the time point, is called long coding. Another example of long coding, which does not involve time, is in family studies where a measurement is made on each member of a family. In wide coding there would be a measurement variable for each member of the family

family	X_1	X_2	X_3	X_4
1	x	x	x	x

whereas in long coding there would be one variable for the measurement X and one for the place in the family:

family	member	X
1	1	x
1	2	x
1	3	x
1	4	x

For most purposes, when using Stata, long coding is better than wide.

Data on forced expiratory volume (FEV) as a percentage of normal, measured every 3 months over a period of 48 months, are coded long in the file `fevlong.dta`, and wide in `fevwide.dta`. To see the difference

```
. describe using fevlong
. describe using fevwide
```

9.2 Graphing repeated measures

For the FEV data it would be helpful to see a graph of forced expiratory volume by month, for each subject. To obtain the graph for subject 1, for example, try

```
. use fevlong, clear
. sort id month
. twoway connected fev month if id==1
```

Each observation is connected to the next one in the file, so the file needs to be sorted by `id`, then `month` within `id`. To see the graphs for all subjects in group 3, try (expect a long wait here)

```
. twoway connected fev month if grp==3, by(id)
```

This shows a separate graph for each subject. To see the graphs for all subjects in group 3 on the same graph, try

```
. twoway connected fev month if grp==3, connect(ascending)
```

The option `connect(ascending)` tells Stata to connect each point to the next provided that the X-values are ascending – in this case provided that the month is ascending. When the data are sorted by `id` and `month`, this has the effect of joining the points until the month changes back to 0, i.e. until the subject `id` changes. If the data are not correctly sorted the connecting lines in the graph are all over the place. For example, try

```
. generate u=uniform()
. sort u
. twoway connected fev month if grp==3, c(ascending)
```

and you will see a tangle of lines.

9.3   Working at the group level

The graphs above are at the subject level, but it might be interesting to see the plot of mean FEV against `month` by `grp`. This can be done by using the `egen` command

```
. egen mfev=mean(fev), by(grp month)
```

to create the means by group and month. To see the result of the `egen` command, try

```
. sort grp month id
. browse
```

You will see that the new variable `mfev`, which contains the mean FEV by group and month, takes a value for each subject but for subjects in the same group and month these values are the same. To graph the means for the three groups try

```
. twoway connected mfev month, c(ascending)
```

At first sight this is surprising, because we have plotted the graphs using data on the subjects, yet the graphs are at group level. The reason for this is that the same graph is being plotted for each subject in group 3, and similarly for groups 1 and 2, so the graphs appear as one per group.

9.4   Collapsing the data

The `collapse` command can be used to create a new dataset which contains the mean FEV by group and month. Try

```
. use fevlong, clear
. collapse (mean) mfev=fev, by(grp month)
. describe
. list
```

The data are still coded long, but the unit of observation is now the group-month not the subject-month, and there are only 41 records. The syntax of the `collapse` command needs a little explanation: the `(mean) mfev=fev` asks for a new variable called `mfev` which contains the mean of the existing variable `fev`. The `by(grp month)` asks for the means to be calculated by group and month, and for the variables `grp` and `month` to be retained in the collapsed dataset. To graph the group means in the collapsed data try

```
. sort grp month
. twoway connected mfev month, c(ascending)
```

Other possibilities for `collapse` include `(median)`, `(sum)`, `(max)`, `(min)`, etc. For a full list with examples, try `help collapse`.

9.5 Reshaping from long to wide and vice versa

Long coding is more versatile than wide, but with wide coding it is easier to make comparisons between different variables. For example, with family data it would be easier to compare the family members with wide coding than with long coding. Fortunately there is a command which will convert from long to wide, or from wide to long.

The information about `fev` is coded long in `fevlong.dta`, and can be converted to wide coding with the command

```
. use fevlong, clear
. reshape wide fev, i(id) j(month)
. describe
```

The `i( )` part of the command contains the variable which identifies the subject, while the `j( )` part refers to the variable which contains the month in which the measurement took place.[1] The variable `fev`, which is being converted from long to wide, follows the keyword `wide`, and the names of the new variables which make up the columns in the wide coding are made up by combining `fev` with the values of `month` to give `fev0`, `fev3`, etc. After the `reshape` command you will see more variables and fewer records, because each record now corresponds to a subject. You can now convert from wide to long and long to wide, as often as you wish, with

```
. reshape long
. describe
. reshape wide
. describe
```

Now suppose we started with wide coding. The FEV data are also coded wide in the file `fevwide.dta`, where the records look like this:

```
id  grp   fev0   fev3   fev6 ...   fev48
 1    1  60.58  65.35  55.87 ...   65.68
```

To convert to long, try

---

[1] Mathematicians commonly use the symbols i for rows (subjects) and j for columns (variables), which explains why they are used here.

```
. use fevwide, clear
. reshape long fev, i(id) j(month)
. describe
```

The name within the round brackets in j( ) refers to a variable that did not exist
in fevwide but is created for the long format and takes the numerical information
which follows fev in fev0, fev3,fev6, ..., fev48. Try help reshape for further
details and examples.

9.6   Use of system variables with by:

The system variable _n indexes each record in turn, from 1 to _N, where _N is the
total number of records. Try

```
. use fevlong, clear
. count
. display _N
. generate record = _n
. browse record id month
```

and you will see that the variable record takes the value 1 for the first record, 2 for
the second, and so on up to the last. When used with by the variable _n refers to
the records in each group defined by the by, and the variable _N refers to the total
number of records in each group. For example, for the observations in a group of 3
records, defined by a variable grp taking the value 1, the variables _n and _N take
the values:

grp	_n	_N
1	1	3
1	2	3
1	3	3

To illustrate some of the uses of these two system variables, we shall first generate a
new variable visits which records the number of times each subject visits the clinic
for an FEV measurement, try

```
. sort id
. by id: generate visits = _N
. browse
```

For each value of id the variable visits takes as its value the number of records for
that id, i.e. the number of visits. You will see that subject 1 makes 15 visits, and
that the variable visits has been given the value 15 for each of the 15 observations
on subject 1. If you tabulate visits with

```
. tabulate visits
```

the total frequency is 663, the number of records, not the number of subjects, which is 57.

The best way of making the table of values of `visits` refer to subjects is to create the variable only for the first record for each subject. This is done by using `_n` to restrict the `generate` command to the first record:

```
. drop visits
. by id: generate visits = _N if _n == 1
. browse
```

You will see that `visits` takes the value 15 for the first record for subject 1, but is missing for the others. Now try

```
. tabulate visits
```

which tabulates the frequency of visits made by each subject: six subject makes 3 visits, one subject makes 4 visits, etc.

As another example, suppose we wish to create a variable `up` which measures the increase in FEV from 0 to 3 months. To do this, try

```
. sort id month
. by id: generate up = fev[2] - fev[1] if  _n==1
. browse
```

Note the use of `fev[1]` and `fev[2]` to pick out the first and second observations for each subject – for this to work, the data must be sorted by `id` then `month` within `id`.

Another way of doing this is with `bysort`:

```
. drop up
. bysort id (month): generate up = fev[2] - fev[1] if  _n==1
```

The parentheses around `month` indicate that the `sort` is on both `id` and `month` but the `by` refers only to `id`.

9.7 Merging files with long coding

In the examples of merging described in Chapter 8 there were no records with the same identifier within the master file - each had a unique identifier. The same was true for the using file. But this is not the case when repeated measurement data are coded long. For example, in a typical repeated measurement study, data which do not change with time, such as sex and age at entry, are collected in one file, while data which do change with time, such as blood pressure, are collected as one record per visit in another file.

As an example, the file `agesex.dta` contains information for three subjects, and is sorted on the subject identifier `id`:

id	age	sex
100	47	m
101	.	f
102	67	m

while the file `bp.dta` contains data on blood pressure for each visit to a clinic, also sorted on the subject identifier `id`. In this second file, if subject 100 makes 3 visits there will be 3 records with the identifier 100:

id	visit	bp
100	1	180
100	2	160
100	3	155
101	1	160
102	1	120
102	2	140

So the first file holds subject-level information while the second one holds visit-level information. The first file will have as many records as there are subjects, while the second file will hold a larger number of records, one for each visit. Both files are already sorted on `id`, so to merge them, matching on `id`, try

```
. use agesex, clear
. merge id using bp
. tabulate _merge
. sort id visit
. list
```

You will see that the merge has been successful and that there are several records for the same `id`, each containing both subject-level and visit-level information:

id	age	sex	visit	bp	_merge
100	47	m	1	180	3
100	47	m	2	160	3
100	47	m	3	155	3
101	.	f	1	160	3
102	67	m	1	120	3
102	67	m	2	140	3

A matched merge will work provided that in one of the files the records are uniquely identified by the matching variables; if not Stata will still merge the files, but the result may not be what you expect (or want). The rules which Stata follows in this situation are described in the *Stata Reference Manual*, but such merges are best avoided.

Exercises

1. Most countries publish mortality rates, separately for each 5-year age-band, every five years. The file `lungcalong.dta` contains the mortality rates for lung cancer, for males in Finland, coded long. Load this file and list its contents. Each observation corresponds to an age group, and a calendar period, and contains a single rate per 100 000. This is long coding.

2. Graph the rate against period, separately for each age group, using a log scale for the rate. You can do this by using the option `yscale(log)` with `twoway connected`. Give careful thought to how the data should be sorted.

3. Change the marker label to the code for age group by using the option `mlabel(age)`. Label the Y-axis in a better way.

4. Graph the the rate against age group, separately for each period, using a log scale for the rate. Use the code for period as the marker label.

5. The code for period takes up too much space, so create a new code

   ```
   . generate pcode=(period-1970)/5
   ```

   and use `pcode` as the marker label.

6. Most commonly, data of this kind are first coded wide, with the rates for all age-bands included in a single observation. This is illustrated in the file `lungcawide.dta`. Load this file and list its contents.

7. To reshape this as long, try

   ```
   . reshape long rate, i(period) j(age)
   . describe
   ```

   Note that the variable `age` takes the numerical information following `rate` while `period` refers to the current (wide) observations, one for each period.

8. Load the data in `fevwide` and remind yourself of its contents. Use `reshape long` to create a new variable called `ratio` which is the ratio between the first and last FEV measurements for each subject.

Chapter 10

Response and explanatory variables

In this chapter we shall start by considering how variables are to be used in a statistical analysis. The main distinction is between response and explanatory variables. The type of response carries useful information about how to summarize the distribution of the response in tables.

10.1   Questions in statistical analysis

Most questions in statistical analysis take the form of asking whether the value which one variable takes for a given subject depends on the value taken by another variable. For example, in the births data we might be interested in whether the birth weight of a baby depends on whether it is a boy or girl. The variable which is of primary interest, in this case the birth weight, is called the response variable; the variable on which the response variable may depend, in this case the sex of the baby, is called the explanatory variable. In biostatistics four types of response are particularly common:

1. Binary

2. Metric

3. Failure

4. Count

A binary response has just two values which should be coded 0 and 1. A metric response (also called a quantitative response) measures some quantity and usually has many possible values. A failure response indicates whether or not a subject fails at the end of a period of observation, and is used with survival data. Finally, a count response records a number of events, and often arises with aggregated failure data. The type of response determines how it will be summarized. For example, a binary response is usually summarized using the proportion of 1's, and a metric response is usually summarized using its mean or median. The following questions illustrate response and explanatory variables and refer to the births data.

Does the birth weight of a baby depend on whether the mother was hypertensive?

The response variable is `bweight` which is metric and the explanatory variable is `hyp`. Because the response is metric the distribution of response can be summarized using either the mean or the median. Such a table of means, for example, can be produced with the commands

```
. use births, clear
. table hyp, contents(freq mean bweight)
```

Does low birth weight depend on the sex of the baby?

The response variable is now `lowbw` which is binary and the explanatory variable is `sex`. The distribution of `lowbw` by `sex` is given by the relative frequencies of its values, obtained with

```
. tabulate lowbw sex, col
```

which shows that 10% of male babies are of low birth weight, while 14% of female babies are of low birth weight. The `table` command is meant primarily for summarizing a metric response, but it can be used with a binary response provided that this is coded 0/1. Thus

```
. table sex, contents(freq mean lowbw)
```

again shows that 10% of male babies are of low birth weight, while 14% of female babies are of low birth weight. This works because the mean of a variable coded 0/1 is equal to the relative frequency of the 1's.

10.2 Producing tables with tabmore

To help in the preparation of tables for a variety of responses a dialog box called up by the command `db tabmore`[1] invites you first to specify the type of response. Consider again the question of whether `bweight` varies with `hyp` and try

```
. db tabmore
```

---

[1]The command `tabmore` is not part of official Stata, but it is included with the files which come with this book (see Chapter 0). It replaces the commands `tabmenu1` and `tabmenu2` which were available with the Stata 7 version of the book.

to bring up the dialog box.

- Enter `bweight` in the *Response variable* box, select *metric* in the *Type of response* menu, and enter `hyp` in the *Row variable* box.

- Select the *Summary* tab. The response `bweight` is metric, and under metric you are offered three possibilities for summarizing the response - *Mean, Geometric mean,* and *Median*. Select *Mean*, and press *OK* to produce

```
. tabmore, res(bweight) typ(metric) row(hyp) mean
```

The output looks like this:

```
Response variable is: bweight which is metric
Row variable is: hyp
Number of records used:   500


Summary using means
---------------------
hypertens |   bweight
----------+----------
        0 |   3198.90
        1 |   2768.21
---------------------
```

The figures in the table are the mean values of the babies' birth weights for mothers who were normal (`hyp=0`) or hypertensive (`hyp=1`). To get more information you could select the *More output* tab where you can select *Frequencies and Confidence intervals*. Try this now.

With a binary response, such as `lowbw`, the choice of summary is different. Try

```
. db tabmore
```

to bring up the dialog box, and press R to clear previous selections.

- Enter `lowbw` in the *Response variable* box, select *binary* as the *Type of response*, and enter `hyp` in the *Row variable* box.

- Select the *Summary* tab. Because the response is binary you will be offered a choice between *Proportions* and *Odds*. Select *Proportions* and press *OK* to produce

```
. tabmore, res(lowbw) typ(binary) row(hyp) prop
```

The output looks like this:

```
Response variable is: lowbw which is binary
Row variable is: hyp
Number of records used:  500

Summary using proportions per 100
---------------------
hypertens |     lowbw
----------+----------
       0 |      9.35
       1 |     27.78
---------------------
```

The figures in the table are the percentages of low birth weight babies for mothers who were normal or hypertensive. To produce 90% confidence intervals for these proportions, check *Confidence intervals* in the *More output* tab of the dialog box, and enter 90 in the *Level of confidence* box. Press *OK* to produce

```
. tabmore, res(lowbw) typ(binary) row(hyp) prop ci level(90)
```

The command can be edited, so small changes can be made without going through the dialog box again.

10.3   A second explanatory variable

We found a strong relationship between birth weight and whether the mother was hypertensive, but is this relationship the same for both male and female babies? To study this we need to produce a table of mean birth weight by both `hyp` and `sex`:

```
. table hyp sex, contents(freq mean bweight)
```

shows that the birth weight of both male and female babies is lower when the mother is hypertensive than when the mother is normal – about 500 g lower for males babies and about 400 g for female babies.

To produce the same table using `tabmore`, start with

```
. db tabmore
```

and press R to clear previous selections. Then enter `bweight` as *Response variable* and select *metric* as *Type*. Enter `hyp` as *Row variable*, and `sex` as *Column variable*. Select the *Summary* tab, select *Mean*, and press *OK* to produce

```
. tabmore, res(bweight) typ(metric) row(hyp) col(sex) mean
```

To reverse the rows and columns, check the box in the *Main* dialog box marked *Reverse rows and cols*.

10.4   Odds

Odds are less familiar than proportions, but they measure the same thing. When 60 babies out of 500 are low birth weight the proportion is $60/500 = 0.12$, while the odds of being low birth weight are $60/440 = 0.1364$. The interpretation of the proportion 0.12 is that for every 100 babies, 12 are low birth weight; the interpretation of the odds 0.1364 is that for every 100 normal birth weight babies there will be 13.64 low birth weight babies. Odds can be obtained from proportions using

$$\text{Odds} = \frac{\text{Proportion}}{1 - \text{Proportion}} = \frac{0.12}{1 - 0.12} = 0.1364$$

Similarly proportions can be obtained from odds using

$$\text{Proportion} = \frac{\text{Odds}}{1 + \text{Odds}} = \frac{0.1364}{1 + 0.1364} = 0.12$$

Odds are used mainly with case-control studies (see Section 10.5), but also have some technical advantages because they are not constrained to be less than 1, as proportions are. To make a table of the odds of the baby being low birth weight for normal and hypertensive mothers, start with

. db tabmore

and press R to remove previous selections. Enter lowbw as *Response variable*, and select *binary* as *Type*. Enter hyp as *Row variable*, select *Summary* and check *Odds*. Press *OK* to produce

. tabmore, res(lowbw) typ(binary) row(hyp) odds

Tables with too many rows are not particularly useful, and for this reason an upper limit of 10 values for the row and column variables has been set in tabmore. This can be increased in the *Main* tab, if required.

10.5   Case-control studies

Odds are important in case-control studies where, instead of sampling all subjects equally, a different sampling fraction is used for subjects who have a disease (the cases) than for those who do not (the controls). This is called outcome-based sampling in econometrics. As an example, we shall look at a study of physical activity at work and tuberculosis (TB), one of the first case-control studies to be carried out (see Table 10.1). The cases were cases of TB among outpatients at a hospital, and the controls were chosen from outpatients at the same hospital, who were not suffering from TB[5]. Start by listing the data in the file guy.dta with

Table 10.1: Physical activity at work for 1659 outpatients

Activity of physical activity	Tuberculosis (Cases)	Other diseases (Controls)
Little (1)	125	385
Varied (2)	41	136
More (3)	142	630
Great (4)	33	167
Total	341	1318

```
. use guy, clear
. list, sep(0)
. list, sep(0) nolabel
```

The variable `activity` is coded 1, 2, 3, 4 for the four levels of physical activity, and the variable `d` is coded 1 for a case and 0 for a control. The data are aggregated, and the variable `N` contains the frequency with which each combination of `activity` and `d` occurs. One way of dealing with data of this kind is to make it into individual records with the command `expand`:

```
. expand N
. drop N
. summarize
```

The variable `N` is dropped after the expansion because it is no longer relevant. There are now two possible approaches to the analysis (Clayton & Hills, 1993[2]). In the *retrospective* approach we argue from disease back to exposure so the response is `activity`, and the explanatory variable is `d`. The values of `activity` are measuring physical activity on some sort of metric scale, so we shall treat `activity` as metric, and try

```
. tabmore, res(activity) typ(metric) row(d) mean
```

You should see something like

```
Response variable is: activity which is metric
Row variable is: d

Summary using means

        d |     activity
----------+-----------
        0 |       2.44
        1 |       2.24
```

which shows that cases of TB were, on average, less physically active than controls. In the *prospective* approach we argue from exposure forward to disease so the response is d, which is binary, and the explanatory variable is `activity`. Try

```
. tabmore, res(d) typ(binary) row(activity) odds
```

and you should see something like

```
Response variable is: d which is binary
Row variable is: activity


Summary using odds per 100

activity |           d
---------+-----------
  little |       32.47
  varied |       30.15
    more |       22.54
   great |       19.76
---------------------
```

which shows that the odds of being a case decreases with the level of physical activity. Both retrospective and prospective analyses are useful, but on the whole the prospective one is more informative.

Note that the odds which are being tabulated refer to the odds of being a case in the study, not the population. However, it can be shown that

$$\text{Odds in study} = K \times \text{Odds in population,}$$

where K is the ratio between the sampling fractions for cases and controls. So provided the sampling fractions do not depend on `activity`, it follows that if the study odds are going down with `activity`, then the population odds are also going down with `activity`. See Clayton & Hills (1993), p153[2] for a more detailed discussion of this point.

Now you can see why odds are chosen for the analysis of case-control studies: suppose that, instead of 1318 controls, there had been 10 times as many; the odds would now be (apart from random variation),

```
Summary using odds

activity  |          d
----------+----------
   little |      3.247
   varied |      3.015
     more |      2.254
    great |      1.976
```

So although the odds have changed drastically because of the change in sampling the controls, the trend in the odds with `activity` is unchanged. This simple relationship between the odds and the sampling fractions does not hold for proportions.

10.6 Survival data and rates

Data involving survival times are often summarized using *rates*, i.e. the number of events per unit time. There are no survival time variables in the births data, so we need another dataset to demonstrate how to make tables of rates. Try

```
. use diet, clear
. describe
```

These data refer to a follow–up study of 337 male subjects who were asked to weigh the different components of their diet for a week[8]. They were then followed until

1. They developed, and possibly died from, coronary heart disease (CHD)

2. They died from some other cause, or were withdrawn from the study for some reason, or the study ended.

The time for which each subject is followed is the true survival time in the first case, but in the second case the true survival time has been censored. To record data with true and censored survival times we need two variables: the time spent in the study and a variable which indicates whether the subject developed CHD or not. These are called the *follow-up* and *failure* variables, respectively. The response is a combination of these two variables.

In this example the follow-up variable is y, and the failure variable is chd, coded 1 if the subject developed CHD during the period of the study, and 0 otherwise. For a preliminary analysis the total energy intake per day is converted to a binary variable hieng coded 1 if the energy intake is > 2750 kcal, and 0 otherwise. To create a table of rates for chd by hieng, try

```
. db tabmore
```

and press R to remove previous selections. Then enter `chd` as *Response variable*, select *failure* as *Type* and enter y as *Follow-up time*. Enter `hieng` as the *Row variable*, select the *Summary* tab and check *Rates per 1000*. Press *OK* to produce

```
. tabmore, res(chd) typ(failure) row(hieng) rate fup(y)
```

Note that with rates the failure variable is selected as response. Rather unexpectedly, eating a lot seems to prevent CHD (7.07 compared with 13.60). This is because what you eat is largely determined by your physical activity, and a high level of physical activity helps prevent CHD.

10.7   Count data and rates

An example where the response variable is a count arises in the mortality data taken from Rothman (1986)[9]. These data refer to the total deaths and populations in 1962 in Panama and Sweden, by three age categories. Load the data with

```
. use mortality, clear
. describe
. list, sep(0)
```

The data are aggregated, so the response is the total number of deaths in each age x country category, which is in the variable `deaths`. Like failure data, count data also requires a follow-up time, and assuming each subject is followed for the whole of 1962, this is equal to the population multiplied by 1, which is the same as the variable `pop`. To find the mortality rate by nation using `tabmore`, try

```
. db tabmore
```

and press R to remove previous selections. Then enter `deaths` as *Response variable*, select *count* as *Type*, and enter `pop` as *Follow-up time*. Enter `nation` as *Row variable*, select the *Summary* tab, and check *Rates per 1000*. Press *OK* to produce

```
. tabmore, res(deaths) typ(count) row(nation) rate fup(pop)
```

The output shows that the overall (crude) mortality for Sweden is slightly higher than for Panama. A rather more accurate comparison is made by including `agegrp` as the column variable:

```
. tabmore, res(deaths) typ(count) row(nation) col(agegrp) rate fup(pop)
```

Now it is clear that in the first two age groups the mortality for Sweden is lower than the mortality for Panama. Only in the third age group is it higher.

Exercises

1. Start by loading the births data. Babies are classified as pre-term if they are born before 37 weeks. Does being pre-term depend on whether the mother is hypertensive or not? Prepare a table of proportions, with confidence intervals, to investigate this question using `tabmore`.

2. Does this relationship hold for both male and female babies? Investigate this by making a table of the proportion of pre-term babies by `hyp` and `sex`.

3. Does being pre-term change with maternal age? Investigate this by grouping the values of `matage` into a new variable, and preparing a table of proportions of pre–term babies by the categories of this new variable.

4. Repeat the previous table using odds in place of proportions; include 90% confidence intervals.

5. Load the diet data and prepare a table of rates per 1000, and confidence intervals, for `chd` by `job` using `tabmore`.

Chapter 11

Measuring effects

The word effect is a general term referring to ways of comparing the values of the response variable at different levels of an explanatory variable. This chapter covers the measurement of effects as

- Differences in means for a metric response.

- Differences in proportions for a binary response.

- Ratios of odds or proportions for a binary response.

- Ratios of rates for a failure response.

11.1   A metric response

In Chapter 10 we prepared a table of mean `bweight` by `hyp` using `tabmore`. This showed that hypertensive mothers had babies which were on average about 430 g smaller than those delivered to normal mothers. This difference in mean birth weight is called the *effect* of hypertension on birth weight. The command `db effects`, like `db tabmore`, brings up a dialog box which first invites you to specify the type of response.[1] When you bring up this dialog box you will see that the terms *exposure variable* and *stratifying variable* are used instead of *row variable* and *column variable*. The reason for this change is that when calculating effects it is important to distinguish between the different roles which the variables have in an analysis. The variable whose effects you want to calculate is called the *exposure* variable; *stratifying variables* will be introduced later in this chapter. In this example the response variable is `bweight` and the exposure variable is `hyp`, so try

---

[1]The command `effects` is not part of official Stata, but it is included with the files which come with this book (see Chapter 0). It replaces the commands `effmenu1` and `effmenu2` which were available with the Stata 7 version of the book.

```
. use births, clear
. db effects
```

then press R to clear previous selections.

- Enter `bweight` as *Response variable* and select *metric* as *Type*.

- Enter `hyp` as *Exposure*, and notice, at the bottom of the dialog box, that the exposure variable is treated as categorical by default.

- Select the *Measures of effect* tab, and check *Difference in means*. Press *OK* to produce

```
. effects, res(bweight) typ(metric) exp(hyp) exc md
```

The option `exc` states that the exposure is categorical. You will see that the effect of `hyp` on `bweight`, measured using the difference in the mean response, is −430.7 g. The 95% confidence interval for this effect is from −586 to −276 g. The statistical test is for the null hypothesis that the true effect of hypertension is zero, and is on 1 degree of freedom (df) because one effect is estimated. The p-value is very low so there is strong evidence against the null hypothesis. Tests of hypotheses are dealt with in more detail in Chapter 14.

Now cut `gestwks` into 4 groups with

```
. egen gest4=cut(gestwks), at(20,35,37,39,45)
. tabulate gest4
```

When comparing the mean birth weight between the four different levels of `gest4` there will be three effects: the effect comparing level 2 with level 1; level 3 with level 1; and level 4 with level 1. The level with which each of the other levels is compared is called the *baseline* (level 1 in this case). To prepare a table of the three effects of `gest4`, try

```
. db effects
```

then enter `gest4` as *Exposure* in place of `hyp`, and press *OK* to produce

```
. effects, res(bweight) typ(metric) exp(gest4) exc md
```

There are three effects, measured as differences in means, and the statistical test is for the null hypothesis that the true values of these effects are all zero is on 3 df because 3 effects are tested. The *Main* dialog allows you to change the baseline from its default value of 1. Try changing the baseline to 3: each of the levels 1, 2, 4 is now compared with level 3. This will add the option `base(3)` to the `effects` command.

11.2   A binary response

When examining the proportion of low birth weight babies according to the length
of their gestation, using gestation time in four groups, each level of gest4 can be
compared with the baseline level using the difference in proportions, or the ratio of
proportions, or the odds ratio. The preferred method is to use the odds ratio.

Start by using tabmore to prepare a table of the odds that a baby has low birth
weight, by the levels of gest4. Now try

. db effects

and press R to clear previous selections. Then enter lowbw as *Response variable* and
select binary as *Type*. Enter gest4 as *Exposure*, select the *Measures of effect* tab, and
check *Odds ratio*. Press *OK* to produce

. effects, res(lowbw) typ(binary) exp(gest4) exc or

You should see this table of three effects comparing levels 2, 3, 4 against level 1, using
odds ratios:

```
Levels        Effect      95% Confidence Interval

2/1            0.164     [  0.05 ,   0.51 ]
3/1            0.029     [  0.01 ,   0.08 ]
4/1            0.003     [  0.00 ,   0.01 ]


Test for no effects

chi2(  3)       =   82.449
P-value         =    0.000
```

Compared with level 1 of gest4, mothers at level 2 have lower odds of a low birth
weight baby by a factor of 0.164, mothers at level 3 have lower odds by a factor of
0.029, while mothers at level 4 have lower odds by a factor of 0.003. The statistical
test that appears below the table is for the null hypothesis that the true values of
these three effects (odds ratios) are all 1. It takes the form of a chi–squared statistic,
and is on 3 df because there are three effects being tested.

This makes more sense if you choose level 4 as the baseline. The effects are then

```
Levels        Effect      95% Confidence Interval

1/4          356.944     [ 84.11 , 1514.78 ]
2/4           58.614     [ 15.36 ,  223.64 ]
3/4           10.349     [  3.00 ,   35.72 ]
```

Compared to women with gestation period greater than or equal to 39 weeks, the
odds of having a low birth weight baby increase by a factor of 10.3 for women with

gestation period in the range 37–39, by a factor of 58.6 for 35–37 and a factor of 357 for women with a gestation period less than 35 weeks. We have chosen to measure the effects of `gest4` as odds ratios, but they can also be measured using the ratios of proportions. Try

```
. db effects
```

select the *Measures of effect* tab, and check *Ratio of proportions*. You will see a table showing the three effects of `gest4` as ratios of proportions instead of odds ratios. You should be aware that when using the ratio or difference of proportions to measure effects, the program may fail to reach an answer for datasets where some of the proportions being compared are close to 0 or 1.

11.3   Case-control studies

For the measurement of effects in a prospective analysis of unmatched case-control studies it is essential to select odds ratios. Try

```
. use guy, clear
. expand N
. drop N
. db effects
```

and press R to clear previous selections. Then enter d as *Response variable* and select binary as *Type*. Enter `activity` as *Exposure*, select the *Measures of effect* tab, and check *Odds ratio*. Press *OK* to produce

```
. effects, res(d) typ(binary) exp(activity) exc or
```

You should see a table of odds ratios comparing levels 2/1, 3/1, and 4/1.

11.4   A failure response

Load the diet data and cut `energy` into 3 groups with

```
. use diet, clear
. egen eng3=cut(energy),at(1500,2500,3000,4500)
. tabulate eng3
```

Use `tabmore` to prepare a table showing the rates of CHD for different levels of `eng3`, and note that the rate goes down from 16.90 to 4.88 per 1000, with increasing levels of `eng3`. The two effects of `eng3` can be measured as rate differences or rate ratios. To prepare a table of effects using rate ratios, try

```
. db effects
```

and press R to clear previous selections. Then enter `chd` as *Response variable*, select failure as *Type*, and enter `y` as *Follow-up time*. Enter `eng3` as *Exposure*, select the *Measures of effect* tab, and check *Rate ratio*. Press *OK* to produce

```
. effects, res(chd) typ(failure) exp(eng3) exc rr fup(y)
```

You should see something like this:

```
Levels          Effect     95% Confidence Interval

2/1             0.645    [  0.34 ,  1.23 ]
3/1             0.289    [  0.12 ,  0.67 ]


Test for no effects

chi2(  2)     =     8.241
P-value       =     0.016
```

Compared with level 1 of `eng3`, subjects at level 2 have a lower rate of CHD by a factor of 0.645, and subjects at level 3 have a lower rate of CHD by a factor of 0.289. The statistical test that appears below the table is for the null hypothesis that the true values of both effects of `eng3` are 1. The p-value is very small, so there is strong evidence against the null hypothesis.

11.5 Metric exposure variables

When using `tabmore` no distinction is drawn between categorical and metric explanatory variables. If a metric variable such as `gestwks` in the births data is selected as explanatory, `tabmore` simply refuses to make a table because `gestwks` has too many values. When using `effects` the distinction between categorical and metric is essential because, as we shall now see, it is possible to find the effects of a metric exposure even when it has many values.

We shall illustrate this by finding the effect of `gestwks` on `bweight`. Even though `gestwks` has many values we can still find the effect of one unit increase in `gestwks` on `bweight` by making the assumption that this effect is the same throughout the range, i.e. that the effect of changing from 30 to 31 weeks of gestation is the same as changing from 31 to 32, and so on throughout the range. If this is the case then the relationship between `bweight` and `gestwks` is *linear*, so start by checking this with

```
. use births, clear
. twoway scatter bweight gestwks
```

You will see that birth weight tends to go up with gestational age, and that the increase in birth weight per unit increase in gestation (in weeks) is roughly constant throughout its range of values. To find the effect of a unit increase in `gestwks`, try

```
. db effects
```

and press R. Then select `bweight` as *Response variable* and metric as *Type*. Select `gestwks` as *Exposure* and check the *Metric exposure* box. Select the *Measures of effect* tab and check *Difference in means*. You are offered a choice of effects *per something*, but the default is per 1 unit, so go with this and press *OK* to produce

```
. effects, res(bweight) typ(metric) exp(gestwks) exm md
```

There is now only one effect, namely 197.0 g per unit increase in `gestwks`, i.e. per week of gestation.

We can do the same thing to find the effect of `gestwks` on `lowbw`, using odds ratios to measure the effect. We are now making the assumption that the odds of a low birth weight baby has is reduced by the same factor for a change in gestation from 30 to 31 weeks, 31 to 32 weeks, and so throughout the range. This is the same as saying that the relationship between the log odds and `gestwks` is linear. The easiest way to check this is to cut `gestwks` into equally spaced groups[2] with

```
. egen gest4=cut(gestwks), at(30(5)45)
```

To look at how the odds change from one group to the next, try

```
. tabmore, res(lowbw) typ(binary) row(gest4) odds graph log
```

You can produce this command with the `tabmore` dialog box if you prefer. You will see that the plot of logodds versus `gest4` is remarkably close to linear, and the assumption of a log-linear relationship between the odds of low birth weight and gestation period is a reasonable one. To find the effect of `gestwks` based on this assumption, try

```
. db effects
```

and press R. Then enter `lowbw` as *Response*, and select binary as *Type*. Enter `gestwks` as *Exposure* and check the *Metric exposure* box. Select the *Measures of effect* tab and check *Odds ratio*. Press *OK* to produce

```
. effects, res(lowbw) typ(binary) exp(gestwks) exm or
```

The effect of a unit increase in `gestwks` is to multiply the odds that the baby has low birth weight by 0.48, i.e. a reduction to 41% of its current level for each extra week of gestation.

We shall now return to the diet data and find the effect of `energy` on the rate of CHD, where `energy` is a metric exposure variable, and the effect is measured as a rate ratio per unit of energy. As before we assume that this effect is the same throughout the range, that is the effect of changing from 1700 to 1701 kcal is the

---

[2]We have omitted 5 mothers with less than 30 weeks gestation to avoid groups in which the odds of low birth weight are infinite.

same as changing from 1702 to 1703, and so on throughout the range. This is the same as saying that the relationship between the log rate and energy is linear. First we check this assumption by cutting energy into equally spaced groups[3], and looking at how the rates change from one level to the next:

```
. use diet, clear
. egen eng5=cut(energy),at(1500(500)3500)
. tabmore, res(chd) typ(failure) row(eng5) rate fup(y) graph log
```

The plot looks reasonably close to linear, so try

```
. db effects
```

and press R. Then enter chd as *Response variable*, select failure as *Type*, and enter y as *Follow-up time*. Enter energy as *Exposure*, and check the *Metric exposure* box. Select the *Measures of effect* tab and check *Rate ratio*. Press *OK* to produce

```
. effects, res(chd) typ(failure) exp(energy) exm rr fup(y)
```

The effect of a unit increase in energy is 0.999, i.e. the CHD rate is reduced by a factor of 0.999 for each increase of 1 kcal in total energy. An increase of 1 kcal is a very small amount of energy, which explains why the effect is so close to 1. It would be better to measure the effect per 100 kcal, or even per 500 kcal. Recall the dialog box by typing db effects and then change *per unit* (bottom of the dialog box, opposite *Metric*) to 100. The effect is now 0.8913 per 100 kcal.

11.6   Metric versus grouped

For a metric exposure it is usually best to group its values and to find the effects by comparing each level of the grouped exposure with the baseline. With small amounts of data there is some advantage in using the exposure in its metric form, and finding the effect per unit of exposure, but the problem is that with small datasets it is not easy to check the assumption of a constant effect throughout the range.

---

[3]The cut is chosen to omit 16 men whose energy intake is above 3500, thus avoiding a group with zero rate.

Exercises

1. Load the births dataset and create the variable `agegrp` with

   ```
   . use births, clear
   . egen agegrp=cut(matage), at(20,30,35,40,45)
   ```

   Using `effects`, find the effects of `agegrp` on `bweight` as differences in means.

2. Repeat the previous question, changing the baseline to level 4.

3. Use `twoway scatter` to check whether birth weight changes linearly with maternal age.

4. Use `effects` to find the effect of `matage` on `bweight`, per year of maternal age.

5. Find the effects of `agegrp` on `preterm` using odds ratios.

6. Use `tabmore` to check on log linearity between `lowbw` and `matage`.

7. Use `effects` to find the effect of `matage` on `lowbw`, per year of maternal age.

Chapter 12

Stratifying and controlling

In this chapter we show how to control for potentially confounding variables by stratifying, estimating the effect(s) of an exposure within strata, and then combining the separate estimates.

12.1   Stratification

The effects calculated so far are *overall* (or crude) effects, and take no account of the possibility of confounding due to other variables. For example, in the `births` data, the overall effect of `hyp` on birth weight is -430.7 g, but the sex of the baby is associated with its birth weight, so if sex is also associated with hypertension, then part of the overall effect could be due to sex differences in the babies of normal and hypertensive women. To exclude this possibility we need to *stratify* the effect of `hyp` for `sex` by selecting `sex` as a stratifying variable. Start with

```
. use births, clear
. db effects
```

and press R. Then enter `bweight` as *Response* and select metric as *Type*. Enter `sex` as the *Stratifying variable*. Enter `hyp` as *Exposure*, select the *Measure of effect* tab, check *Difference in means*, and press *OK* to produce

```
. effects, res(bweight) typ(metric) exp(hyp) str(sex) exc md
Level of sex                  Effect    95% Confidence Interval
1                           -496.3513     [ -296.143 , -696.560 ]
2                           -379.7734     [ -141.606 , -617.941 ]

Overall test for effect modification
chi2( 1)     =    0.542
P-value      =    0.462
```

The effect of `hyp` is -496.3 g for boys and -379.8 g for girls, and no part of these stratified effects can be due to sex differences between the babies of hypertensive and normal women, because the first is based only on boys, and the second only on girls. The p-value refers to the test of the null hypothesis that these two stratified effects are equal, i.e. that the difference between the two effects is zero. It has one degree of freedom because one difference is being tested.

12.2 Controlling

These separate estimates are each based on a subset of the data and have lower precision than the estimate based on the total data, but because there is no evidence that the true values of these two effects differ, they can be combined to give a single effect, -448.08 g, based on all of the data. The way they are combined need not concern us here, but is described in Chapters 13 and 15. This combined effect is called the effect of `hyp` controlled for `sex`. To see how to control the effect of `hyp` for `sex` start with

. db effects

but don't press R. Then remove `sex` as the *Stratifying variable*, select the *Control variables* tab, and enter `sex` as a *Categorical control variable*. Press *OK* to produce

. effects, res(bweight) typ(metric) exp(hyp) catcon(sex) exc md

The result should now look like this

```
Effect(s) of hyp

Levels       Effect     95% Confidence Interval
2/1          -448.08    [  -600.9 ,  -295.3 ]


Test for no effects
chi2(  1)      =   33.030
P-value        =    0.000
```

The effect of `hyp` controlled for `sex` is -448.08 g, and the p-value refers to the test of the null hypothesis that the effect of `hyp` controlled for `sex` is zero.

Although many people control for a potential confounder without first looking at the separate effects at different levels of the confounder, it is better to look first, because the separate effects are combined on the assumption that their true values are the same. The statistical test used to check this assumption is called the test for effect modification. When there is strong effect modification the separate effect of exposure should be reported.

As another example, we shall return to the diet data, and control the effect of `hieng` on the rate of CHD for `job`, first looking at the separate effects of `hieng` in the different jobs. Start by finding the overall effect of `hieng` (level 2 vs level 1) on `chd`, as a rate ratio, with

```
. use diet, clear
. db effects
```

Press R, then use the dialog box to produce

```
. effects, res(chd) typ(failure) exp(hieng) exc rr fup(y)
```

The answer should be 0.520. Now go back to `db effects`, select `job` as the *Stratifying variable*, and press *OK* to produce

```
. effects, res(chd) typ(failure) exp(hieng) str(job) exc rr fup(y)
```

You will now see three effects of `hieng`, one for each level of `job`:

```
Level of job     Effect    95% Confidence Interval

driver           0.410     [ 0.124 , 1.362 ]
conductor        0.655     [ 0.227 , 1.888 ]
bank             0.518     [ 0.212 , 1.267 ]


Overall test for effect modification
chi2(  2)     =    0.331
P-value       =    0.847
```

All three effects are measuring the effect of `hieng`, but in subjects who have different jobs. These three effects are similar in size, so there is no evidence that `job` modifies the size of the effect of `hieng`. This is confirmed by the significance test for effect modification which appears below the table. The chi–squared statistic is on 2 df because we are making two comparisons: 0.655 (conductors) with 0.410 (driver) and 0.518 (bank) with 0.410 (driver). The p-value is large, confirming that there is no evidence that `job` modifies the size of the effect of `hieng`.

We can now combine the three separate effects as a single effect, by removing `job` from being a stratifying variable in the *Main* tab, and selecting `job` as a control variable in the *Control variables* tab. Try this now – you should get 0.525 for the effect of `hieng` controlled for `job`, not very different from the overall, or uncontrolled effect of 0.520, showing that the confounding influence of `job`, if any, is minimal.

It is also possible to control for one variable while stratifying by another. For example, try

```
. db effects
```

but don't press R. Then increase the *Max number of values a categorical control variable can take* to 12. Select the *Control variables* tab and enter `month` as a *Categorical control variable*. Press *OK* to produce

```
. effects, res(chd) typ(failure) exp(hieng) str(job) exc catcon(month)
      maxval(12) rr fup(y)
```

The output shows the effect of `hieng`, controlled for month, but separately by job. Any number of control variables can be selected with `db effects`, but there can be only one stratifying variable.

12.3   Controlling the effect of a metric exposure

The effect of a metric exposure variable can be controlled for a confounding variable, in the same way as for a categorical exposure. For example, try

```
. db effects
```

and press R. Then enter `energy` as *Exposure* and check *Metric exposure*. Enter 100 in the *Per unit* box. Enter `job` as the *Stratifying variable*, select the *Measure of effect* tab, and check *Rate ratio*. Press *OK* to produce

```
. effects, res(chd) typ(failure) exp(energy) str(job) exm rr fup(y)
```

The three effects you will see are the effects of energy per 100 kcal at each of the levels of `job`. They seem similar, and this is confirmed by the test for effect modification. Combine them by moving `job` from being a stratifying variable to being a control variable, to get 0.892 as the effect of `energy` per 100 kcal, controlled for `job`.

12.4   Metric control variables

To complete the story we need to discuss metric control variables. For example, suppose we want to control the effect of `hieng` on `chd` for `height`. On the assumption that the log rate changes linearly with `height`, this is easily done. Start with

```
. db effects
```

and press R. Then enter `chd` as *Response*, select *failure* as *Type*, and enter y as *Follow-up time*. Enter `hieng` as *Exposure*, select the *Measures of effect* tab and check *Rate ratio*. Select the *Control variables* tab and enter `height` as a *Metric control variable*. Press *OK* to produce

```
. effects, res(chd) typ(failure) exp(hieng) exc metcon(height) rr fup(y)
```

The effect of `hieng` controlled for `height` (metric) is 0.613. Of course, before doing this, you should check the linear relationship between the log rate and `height` by grouping `height`, and making a table of rates by grouped height, as follows:

```
. egen htgrp=cut(height),at(150(5)180)
. tabmore, res(chd) typ(failure) row(htgrp) rate fup(y) graph log
```

You will see that the assumption of a linear relationship between the log rate and `height` is not unreasonable.

12.5   Metric versus grouped

For a metric control variable, the issue is whether to control the effects of an exposure using the control variable in its metric form, or whether to group its values and control for it in a categorical form. Very similar results are usually obtained both ways, so this is a less important decision than how to measure the effects of an exposure (metric or categorical) in the first place. In general, when you have lots of data, group the control variable; when you have only a small amount of data, treat the control variable as metric.

When stratifying in order to check for effect modification, there is a strong case for treating the exposure as metric whenever possible. This is because with a metric exposure there will only be one effect per stratum, but when an exposure is grouped with, say, 5 levels there will 4 effects per stratum. Since there is usually only a small amount of data with which to estimate effects in each stratum, the fewer effects to estimate the better.

Exercises

1. Load the births data and find the effect of `hyp` on `preterm` using odds ratios.

2. Is this effect modified by `sex`? If not, find the effect per controlled for `sex`.

3. Find the effect per year of `matage` on `preterm` using odds ratios.

4. Is this effect modified by `sex`? If not, find the effect per year of `matage` controlled for `sex`.

5. Load the mortality data and find the effect of `nation` on the death rate as a rate ratio.

6. Is this effect modified by `agegrp`?

Chapter 13

Regression commands

The command `effects`, introduced in Chapter 11, is a convenient way of working, particularly when starting, but for more advanced work it is necessary to access the regression commands on which it depends. This chapter covers

- The `regress` command for a metric response.

- The `logit`, `logistic`, and `binreg` commands for a binary response.

- The `poisson` command for survival data.

- The `clogit` command for matched case-control studies.

These commands can all be found under the *Statistics* tab.

13.1   Three important regression models

The original (ordinary) regression model was

$$\mu = \alpha + \beta X$$

where μ is the mean of a metric response and X is an exposure. Later this was extended to the logistic regression model for a binary response

$$\ln \left(\frac{\pi}{1 - \pi} \right) = \alpha + \beta X$$

where π and $1 - \pi$ are the probabilities of the two possible responses. The third important model is Poisson regression for a failure or count response

$$\ln(\lambda) = \alpha + \beta X$$

where λ is the rate at which failures or events occur. In each case the variable X is a metric exposure variable, and α is the baseline corresponding to $X = 0$. The parameter β is the effect of a unit change in X on the mean (ordinary regression), or the log odds (logistic regression) or the log rate (Poisson regression).

Although the variable X is always metric in a regression model, there is a simple way of extending the basic model to cover categorical exposures. We shall now give examples of the use of each of the main regression models, starting with a metric exposure, then a categorical exposure with two levels, and finally a categorical exposure with more than two levels.

13.2   A metric exposure

As an example of a metric exposure we shall find the effect of gestation period on birth weight using ordinary regression. The assumption of a constant effect of exposure throughout the range implies a linear relationship between mean birth weight and gestation period of the form

$$\text{Mean birth weight} = \alpha + \beta(X - 24)$$

where for convenience the gestation period (X) has been measured from 24 weeks. Graphically the model looks like this:

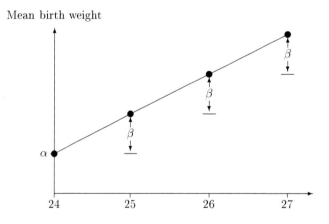

The parameter α is the baseline birth weight at 24 weeks, and β is the increase in the mean birth weight for each unit increase in gestation period. To estimate the values of α and β using Stata, try

```
. use births, clear
. generate gestwks24=gestwks-24
. regress bweight gestwks24
```

Note that the response `bweight` comes immediately after the command, and is followed by the exposure variable `gestwks24`. The important part of the output is shown below. [1]

```
    bweight |     Coef.     [95% Conf. Interval]
------------+----------------------------------
  gestwks24 |  196.9726     179.7054     214.2399
      _cons |  238.2033     -19.11787    495.5244
```

which shows that β is 197 g, with confidence interval from 180 to 214 g. The `_cons` term is 238, and this is the value of α, the baseline birth weight at 24 weeks. If the gestation period had been measured from 0 rather than 24 weeks, α would have been the result of extrapolating the line back to 0 weeks. Try

```
. regress bweight gestwks
```

and you will see that the `_cons` term is now -4489, although β is unchanged.

With `lowbw` as response the regression model is

$$\text{log odds of low birth weight} = \alpha + \beta(X - 24)$$

and β is the change in the log odds of being low birth weight for an increase of 1 week in gestation. To estimate the values of α and β using Stata, try

```
. logit lowbw gestwks24
```

The value of β is -0.8965 and the exponential of this is the odds ratio 0.408. To get this directly, try

```
. logit lowbw gestwks24, or
```

The effect of `gestwks` (or `gestwks24`) is a reduction in the odds of being low birth weight by a factor of 0.408 for every extra week of gestation.

To demonstrate the Poisson regression model we shall find the effect of `energy` in the diet dataset. The regession model is

$$\text{log rate} = \alpha + \beta X$$

where X now refers to the variable `energy`. The parameter β is the change in log rate for an increase of 1 unit in X, i.e. 1 kcal. The parameter α is the baseline log rate when the energy intake is zero. To estimate the values of α and β using Stata, try

---

[1]The first half of the output, headed Source, describes the partition of the total variability in the data between that explained by the regression model and what is left (the residual). In most applications this is only of historic interest.

```
. use diet, clear
. poisson chd energy, e(y)
```

Note that the follow-up time is given in the option `e(y)`. The value of β is -0.0011507 and the exponential of this is 0.9988. You can get this directly with

```
. poisson chd energy, e(y) irr
```

To get the effect of energy per 100 kcal (say), try

```
. generate energy100=energy/100
. poisson chd energy100, e(y) irr
```

The effect is now a reduction by a factor of 0.8913 per 100 kcals increase in energy.

13.3   A categorical exposure with two levels

Using the births dataset the commands

```
. use births, clear
. table hyp, c(mean bweight)
```

show that the mean birth weight for babies born to normal women is 3198.9 while for hypertensive women it is 2768.2. Using the regression model

$$\text{mean birth weight} = \alpha + \beta X$$

where X now refers to the variable `hyp`, we see that because `hyp` is coded 0 for normal and 1 for hypertensive mothers the model predicts the mean birth weight for the two groups as α (substituting $X = 0$) and $\alpha + \beta$ (substituting $X = 1$). In terms of both data and model

	Data		Model
hyp	Mean birth weight	X	Predicted
0	3198.9	0	α
1	2768.2	1	$\alpha + \beta$

In this very simple case the value of $\alpha = 3198.9$ and

$$\beta = 2768.2 - 3198.9 = -430.7.$$

The parameter β can be interpreted as either the effect on birth weight per unit change in X (metric) or the effect of changing from the normal to the hypertensive group (categorical). If `hyp` were coded 1/2 instead of 0/1 the value of α would change, but not the value of β. Since α is rarely of interest this means that an exposure on two levels can generally be treated as metric even though it is coded 1/2.

To estimate the values of α and β using Stata, try

```
. regress bweight hyp
```

The important part of the output is

```
bweight |    Coef.   [95% Conf. Interval]
--------+-------------------------------
    hyp | -430.6959  -585.821   -275.5707
   _cons |  3198.904  3140.038    3257.77
```

The estimated values of β and α are -430.7 and 3198.9, and the last two columns show the 95% confidence intervals. To change these to 90% try

```
. regress bweight hyp, level(90)
```

or you could use

```
. set level 90
```

which changes the level to 90% until you re-set it, or close Stata.

When the response is `lowbw`, which is binary, the effect of an exposure such as `hyp` is usually measured as an odds ratio. The command

```
. tabmore, res(lowbw) typ(binary) row(hyp) odds
```

shows that the odds of low birth weight per 100 for the two groups are

```
----------------------
hypertens |  response
----------+-----------
        0 |     10.31
        1 |     38.46
----------------------
```

The corresponding logistic regression model is

$$\text{logodds of low birth weight} = \alpha + \beta X$$

where X refers to `hyp`. The model now predicts the logodds of being low birth weight for the two groups as α (normal) and $\alpha + \beta$ (hypertensive). In terms of both data and model

	Data			Model	
hyp	Odds of lowbw (per 100)	Logodds	X	Predicted	
0	10.31/100	$\ln(10.31/100)$	0	α	
1	38.46/100	$\ln(38.46/100)$	1	$\alpha + \beta$	

In this simple case $\alpha = \ln(10.31/100) = -2.272$, and

$$\beta = \ln(38.46/100) - \ln(10.31/100) = \ln(38.46/10.31) = \ln(3.730) = 1.316$$

To do this in Stata, try

```
. logit lowbw hyp
```

and to convert from the log scale, try

```
. logit lowbw hyp, or
```

The command

```
. logistic lowbw hyp
```

does the same thing, but has a few more options. Rather pedantically, neither `logit` with `or`, nor `logistic`, report the baseline value of α because it is not an odds ratio.

The related command `binreg` can be used to calculate effects as ratios of proportions, or differences between proportions. For example, to use the ratio of proportions (risk) instead of odds, try

```
. binreg lowbw hyp, rr
```

where the option `rr` stands for *risk ratio*. To use the difference between proportions, try

```
. binreg lowbw hyp, rd
```

where `rd` stands for *risk difference*. The `binreg` command can also be used for odds ratios by using the option `or` – the results should be the same as with `logit, or`.

For survival time data, when effects are measured as rate ratios, the `poisson` command is used. To find the effect of `hieng` as a rate ratio, using the diet dataset, try

```
. use diet, clear
. poisson chd hieng, e(y) irr
```

The option `irr` asks for incidence rate ratios (i.e. rate ratios).

13.4   Categorical exposures with more than 2 levels

In the births example above the exposure, `hyp`, was categorical with two levels, and because it had only two we could treat it as metric. This does not work in general. Suppose, for example, that X referred to a categorical exposure with 3 levels, coded 1,2,3. Changing from level 1 to level 2 is a unit change in X, as is changing from level 2 to level 3, but there is no reason to assume that the effects of these two changes are the same. To deal with this it is necessary to use *indicator variables*, defined in the table below.

Exposure level	X_1	X_2	X_3
1	1	0	0
2	0	1	0
3	0	0	1

X_1 indicates level 1 by taking the value 1 for subjects whose exposure is at level 1, and 0 for all others. Similarly for X_2 and X_3. Using indicators, the right-hand side of the regression model is

$$\alpha + \beta_1 X_1 + \beta_2 X_2 + \beta_3 X_3$$

Of course there are too many parameters at the moment (3 groups and 4 parameters) so one must be dropped. To make the baseline for the effects equal to level 1 of the exposure, the variable X_1 is removed from the regression to give

$$\alpha + \beta_2 X_2 + \beta_3 X_3$$

Substituting the value of the indicator variables for the different levels of exposure shows that this model reduces to

$$\begin{array}{ll} \alpha & \text{for level 1} \\ \alpha + \beta_2 & \text{for level 2} \\ \alpha + \beta_3 & \text{for level 3} \end{array}$$

so that β_2 is the effect of exposure (level 2/1) and β_3 is the effect (level 3/1).

To show how to do this in Stata we shall cut `gestwks` into 4 categories with

```
. use births, clear
. egen gest4=cut(gestwks), at(20,35,37,39,45)
. table gest4, contents(mean bweight)
```

The `table` command shows that the mean birth weight goes up from one category of gestation time to the next. To make a comparison between the mean birth weights for different categories of gestational age we have to create indicator variables for each of the levels of `gest4`. One way of doing this is with

```
. tabulate gest4, generate(ind)
```

which generates four indicator variables called `ind1` to `ind4`. To see how these are coded try

```
. sort gest4
. browse gest4 ind1-ind4
```

For subjects at the first level of `gest4` the variable `ind1` is coded 1, but the other three are coded 0. For subjects at the second level of `gest4` the variable `ind2` is coded 1, but the other three are coded 0, and so on. To set the first level of `gest4` as the baseline against which the other levels are compared, the first indicator variable is omitted from the regression:

```
. regress bweight ind2 ind3 ind4
```

The coefficients of the indicators are the three effects of gest4, using the first level as the baseline. To make the second level the baseline, omit the second indicator and try

```
. regress bweight ind1 ind3 ind4
```

If all three indicators are included in the regression then Stata will recognize that there are too many parameters, and drop one. Try

```
. regress bweight ind1 ind2 ind3
```

and you will see that Stata has dropped ind1. To drop α, and fit the regression model

$$\beta_1 X_1 + \beta_2 X_2 + \beta_3 X_3 + \beta_4 X_4$$

you can use the option noconstant. The coefficients will now be the group means. For example, try

```
. regress bweight ind1 ind2 ind3 ind4, noconstant
```

and you will see that the coefficients are the mean birth weights for the four gestation groups.

It is not necessary to generate your own indicator variables. Instead the generation of indicator variables can be automated with the xi command, which stands for *expand indicators*. Try

```
. xi: regress bweight i.gest4
```

The xi warns Stata that there are categorical variables in the command that follows, and the i. states which variables are categorical. The command

```
. describe
```

shows that variables with names like _Igest4_35 have been added to the dataset. These are the indicator variables generated by xi. Their names are a combination of the name of the categorical variable and the categories they indicate. Since gest4 had four categories, only three indicators are generated and included in the regression. The lowest category is the one used as baseline. You can change the baseline category with

```
. char gest4[omit] 37
```

which omits the category coded 37, i.e. makes this category the baseline. Try

```
. xi: regress bweight i.gest4
```

again, to see the effect of changing the baseline to the category coded 37. Use

```
. char gest4[omit]
```

to re–set the default baseline.

Returning to `hyp` as the explanatory variable, find the effect of `hyp` on `bweight`, treating `hyp` first as categorical, then as metric, with

```
. xi: regress bweight i.hyp
. regress bweight hyp
```

You will see that the results are the same. Leaving out the `i.` makes Stata treat the variable as metric.

To compare the odds of the baby being low birth weight between groups of gestational age, try

```
. xi: logistic lowbw i.gest4
```

The baseline level (by default) is the first. The effect of changing from the first to the second level is to decrease the odds by a factor of 0.164; from the first to the third the odds are reduced by a factor of 0.029, and from the first to the fourth by a factor of 0.003. To change the baseline to the fourth level, try

```
. char gest4[omit] 39
. xi: logistic lowbw i.gest4
```

13.5   Fitted values and residuals

Following any regression command it is possible to predict the response for each subject, using the regression model. These predictions are called fitted values. The difference between the fitted values and the observed values are called the residuals, and these can be useful when checking on how well a regression model fits the data.

For the births data, try

```
. regress bweight gestwks
. predict bwpred
```

which places the fitted values from the regression of `bweight` on `gestwks` in the new variable `bwpred`. The `predict` command picks up the results left behind by `regress` to do this, and any name can be used for the new variable which contains the predictions. Similarly

```
. predict bwres, r
```

places the residuals from the regression of `bweight` on `gestwks` in the new variable `bwres`. A plot of the residual against the observed values, plus the best fitting straight line, is obtained with

```
. twoway (scatter bwres bwpred) (lfit bwres bwpred)
```

and shows no obvious departure from linearity.

When following `logit` with `predict` there is a choice of saving a variable holding the predicted probability of low birth weight babies (the default) or the predicted log of the odds for being low birth weight. The latter is found with the option `xb`. Plotting it against `gestwks` shows that the predicted log odds are linearly related to the explanatory variable.

```
. logistic lowbw gestwks
. predict lpred, xb
. twoway line lpred gestwks, sort
```

On the probability scale the plot of the predicted probabilities against `gestwks` shows a curved relationship:

```
. predict ppred
. twoway line ppred gestwks
```

Pearson residuals can also be obtained with

```
. predict pres, r
```

13.6   Case-control studies

Unmatched case-control studies can be analysed prospectively with `logit` using as response the variable which records whether or not the subject is a case. For example, try

```
. use guy, clear
. expand N
. drop N
. xi: logit d i.activity, or
```

to see the three effects of `activity` as odds ratios.

Matched case-control studies, where the matching has been on variables such as age and sex, can also be analysed using logistic regression, but you must control for the matching variables by including them in the model. Failure to do so will result in a biased effect of exposure, closer to 1 than it should be. This is an unfortunate consequence of using odds.

For individually matched studies, where the data consist of matched sets containing a case and one or more matched controls, the requirement that the analysis controls for the matched sets means that logistic regression cannot be used, because there would be one parameter for each set, which would be too many for the amount of data. Collecting more data does not help because every new matched sets adds a new parameter. Instead it is necessary to use conditional logistic regression.

To illustrate the analysis of matched case-control studies we shall use data from a study of an outbreak of salmonellosis in Denmark[7]. In the Autumn of 1996 an unusually large number of *Salmonella* Typhimurium cases were recorded in Funen County in Denmark. The Danish Zoonosis Centre set up a matched case-control study to find the sources. Cases and two age- sex- and residency-matched controls were telephone interviewed and asked whether they had travelled abroad, or eaten any of a list of foods, in the previous two weeks. The list included beef, pork, veal, poultry, liverpaste, vegetables, fruit, and eggs. The participants were also asked at which retailer(s) they had purchased meat, and independently of this, retailers were linked to meat processing plants. The participants were thus linked to plants, and the results for plant number 7 are of interest here. The data are in the file `salmonella.dta`: the variable `case` identifies cases and controls, and the variable `set` identifies the matched sets. All questions are coded 1 for yes, 0 for no. Load the data and look at the effects of `plant7` with

```
. use salmonella, clear
. clogit case plant7, group(set) or
```

Here the response is `case`, the exposure is `plant7`, and the variable which identifies the matched sets is `set`. You should see that the odds of being a case are 4.47 times higher for subjects who have eaten meat from plant 7 in the last two weeks than for those who have not.

Exercises

1. Load the births data and create the categorical variable `mage4` from `matage` so that it has 4 similarly sized groups. Use `table` (or `tabmore`) to find the mean birth weight of babies born to mothers in these age groups.

2. Use `regress` with `xi` to find the effects of `mage4` on birth weight, with confidence intervals.

3. Repeat the previous question using the metric variable `matage` and find the effect per year of `matage` on birth weight.

4. Use `logistic` to find the effects of `mage4` on the odds of a baby having low birth weight.

5. Repeat the previous question using the metric variable `matage` and find the effect of `matage` per 10 years.

Chapter 14

Tests of hypotheses

In this chapter we introduce the likelihood ratio test and its quadratic approximation, the Wald test. The Stata commands are `lrtest` and `testparm`.

14.1 Models and Likelihood

The fundamental statistical test is the likelihood ratio test. To understand this requires a knowledge of likelihood, and this in turn requires a knowledge of probability models. Some readers might find these concepts unfamiliar, so we shall make a little detour from the main business of the book, and describe them in this section.

Consider a binary response with two outcomes, A and B, and suppose that the observations were

$$\text{A \quad A \quad B \quad A \quad B \quad B \quad B \quad A \quad B \quad B}$$

i.e. 4 A's and 6 B's. As it stands this statement has no scientific interest, because it refers only to the 10 subjects studied. To be of scientific interest it must be generalized to some population of subjects, for example the population of all subjects who could have been in the study. We also need a further assumption, namely that the data arose from the population by a random mechanism. The most common mechanism is random sampling, which is a special case of the binary probability model in which, for each subject, outcome A occurs with probability π and outcome B occurs with probability $1 - \pi$. The probability π is called the parameter of the model, and is the same for all subjects. In the case of random sampling π is the true proportion of subjects with outcome A in the population from which the sample was drawn. Based on the 10 subjects in the study, the estimated value of π is $4/10 = 0.4$.

Likelihood is the probability of the observed data given the probability model which gave rise to these data. It is used to compare different possible values for the parameter of the model; the greater the probability of the data, the more *likely* it is

116

that the parameter value gave rise to the data. In this case the likelihood is

$$
\begin{aligned}
L(\pi) &= \pi \times \pi \times (1-\pi) \times \pi \times (1-\pi) \times (1-\pi) \times (1-\pi) \times \pi \times (1-\pi) \times (1-\pi) \\
&= \pi^4(1-\pi)^6
\end{aligned}
$$

and this expression is used to calculate the likelihood for different values of π. For example, the likelihood for $\pi = 0.2$ is

$$
L(0.2) = (0.2)^4 \times (0.8)^6 = 41.94 \times 10^{-5}
$$

and the likelihood for $\pi = 0.4$ is

$$
L = (0.4)^4 \times (0.6)^6 = 119.44 \times 10^{-5}
$$

From this we can say that $\pi = 0.4$ is more likely to have given rise to the data than $\pi = 0.2$. In this example $\pi = 0.4$ is the value most likely to have given rise to the data – the likelihood for $\pi = 0.4$ is larger than for any other value of π.

To test whether $\pi = 0.2$ could have given rise to the data the likelihood for $\pi = 0.2$ is compared to the likelihood for $\pi = 0.4$ (the most likely value) using the likelihood ratio

$$
L(0.2)/L(0.4) = 41.94/119.44 = 0.3512
$$

The p-value for $\pi = 0.2$ is the probability of observing a likelihood ratio less 0.3512; if the p-value is low then $\pi = 0.2$ could not have given rise to the data. The p-value can be found using a central result in statistics which in this case states that in repeated samples from the model with $\pi = 0.2$, minus twice the natural logarithm of the likelihood ratio has a chi-squared distribution on 1 df. The p-value is therefore the same as the probability of observing a value greater than

$$
-2 \times \ln(0.3512) = 2.0930
$$

in a chi-squared distribution on 1 df. The Stata command

```
. display chi2tail(1,2.0930)
0.14797
```

shows that this probability is 0.148, so there is no evidence against $\pi = 0.2$ and we conclude that this value of π could have given rise to the data. This is called the likelihood ratio (LR) test.

To see whether $\pi = 0.1$ could have given rise to the data, we calculate the likelihood for $\pi = 0.1$, which is

$$
L(0.1) = 0.1^4 \times 0.9^6 = 5.31 \times 10^{-5}
$$

and obtain the likelihood ratio

$$
L(0.1)/L(0.4) = 5.31/119.44 = 0.0445
$$

Minus twice the log likelihood ratio is

$$-2 \times \ln(0.0445) = -2 \times -3.1124 = 6.2245$$

and

```
. display chi2tail(1,6.2245)
0.0126
```

shows that the p-value is 0.0126. Thus the evidence against $\pi = 0.1$ having given rise to the data is quite strong.

14.2   Log likelihood

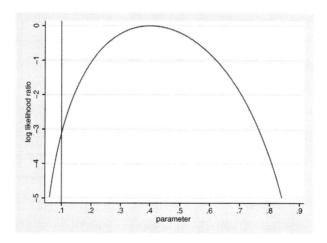

Figure 14.1: Plot of log likelihood ratio against the probability parameter

A more convenient way of looking at a likelihood ratio is to look at the difference in log likelihoods. Writing LL for log likelihood

$$
\begin{aligned}
\text{L}(0.1) &= 5.31 \times 10^{-5}, & \text{LL}(0.1) &= -9.8425 \\
\text{L}(0.4) &= 119.44 \times 10^{-5}, & \text{LL}(0.4) &= -6.7301
\end{aligned}
$$

The log likelihood ratio (LLR) comparing $\pi = 0.1$ with $\pi = 0.4$ is the difference between the log likelihoods,

$$\text{LLR}(0.1) = \text{LL}(0.1) - \text{LL}(0.4) = -9.8425 - (-6.7301) = -3.1123.$$

Figure 14.1 shows a plot of the log likelihood ratio (LLR) comparing different values of π with the most likely value 0.4. You will see that the LLR peaks at $\pi = 0.4$ taking the value zero, and drops to -3.11 when $\pi = 0.1$. If the LLR is plotted against the logodds parameter, the resulting curve is usually a bit closer to a quadratic shape (upside down pudding bowl) – see Figure 14.2. A quadratic approximation to the LLR curve plotted against the logodds parameter is shown in Figure 14.3. The drop in the approximate LLR at $\pi = 0.1$, (logodds equal to $\ln(0.1/0.9) = -2.20$), is -3.80, and the corresponding p-value for $\pi = 0.1$ is found from the probability of values above $-2 \times -3.80 = 7.60$ in a chi-squared distribution on 1 df. When the quadratic approximation is used for the drop in LLR the test is called a Wald test.

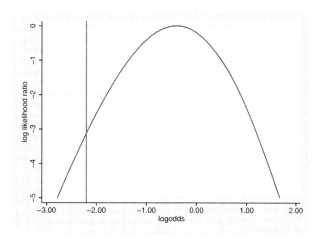

Figure 14.2: Plot of log likelihood against the log odds parameter

14.3   Likelihood ratio and Wald tests in Stata

In practice statistical tests are usually about special values of a parameter which correspond to nothing going on - the so-called null values. For example, if the parameter is an odds ratio then the null value is 1; if it is the log odds ratio, then the null value is the log of 1 which is 0. As an example we shall test whether the odds ratio measuring the effect of maternal hypertension on low birth weight is 1. The effect of hyp on a log scale is found with

```
. use births, clear
. logit lowbw hyp
lowbw |      Coef.    Std. Err.
------+----------------------
  hyp |   1.316614      .31114
_cons |  -2.272126    .1660642
```

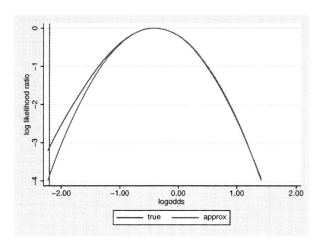

Figure 14.3: Quadratic approximation to the plot of log likelihood against the log odds parameter

so the most likely value of the odds ratio parameter is $\exp(1.3166) = 3.73$. The log likelihood, reported by `logit`, is -175.42. To test the null value of 1 (0 for the log odds ratio) we need to see how much lower the log likelihood is when the parameter takes its null value. This is found by leaving out `hyp` as in

```
. logit lowbw
------------------------------
lowbw |     Coef.    Std. Err.
------+-----------------------
_cons |   -1.99243    .1376205
```

which reports a log likelihood of -183.46. The drop in log likelihood is therefore

$$-183.46 + 175.42 = -8.04$$

and minus twice this is 16.08. The probability of exceeding 16.08 in a chi-squared distribution on 1 df is found from

```
. display chi2tail(1,16.08)
0.00006
```

so the evidence against a log odds ratio of 0 (odds ratio of 1) having given rise to these data is very strong. All this can be automated using the `lrtest` command:

```
. logit lowbw hyp
. estimates store tmp
```

```
. logit lowbw
. lrtest tmp
likelihood-ratio test          LR chi2(1)  =       16.08
(Assumption: . nested in tmp) Prob > chi2 =      0.0001
```

which fits the model including `hyp`, stores the results in somewhere named `tmp` (or whatever name you choose), refits leaving out `hyp` and finally carries out the likelihood ratio test. The last line reminds you that the second model fitted must be obtained from the first by setting some parameters to their null values, i.e. by leaving some variables out.

The Wald test is rather easier to carry out as the model needs to be fitted only once. The command is `testparm`. Try

```
. logit lowbw hyp
. testparm hyp
 ( 1)  hyp = 0

  chi2(  1) =    17.91
Prob > chi2 =     0.0000
```

The chi-squared of 17.91 is close to the value 16.08 obtained with the LR test. The difference is due to the fact that a quadratic approximation has been used for the LLR in the Wald test. Both tests can be used with `logistic` in place of `logit`.

14.4 Joint tests of several parameters

For a categorical exposure on (say) four levels, there are three effects, and it rarely makes sense to test each effect separately. Instead the three null values are tested jointly with a chi-squared on three df. As an example we shall consider whether low birth weight depends on maternal age. Start by cutting maternal age into four groups with

```
. egen agegrp=cut(matage), at(20,30,35,40,45)
. tabulate agegrp
```

and try

```
. xi: logistic lowbw i.agegrp
. estimates store tmp
. logistic lowbw
. lrtest tmp
likelihood-ratio test          LR chi2(3)  =        1.63
(Assumption: . nested in tmp)  Prob > chi2 =      0.6532
```

The test is a test of the null hypothesis that all three true effects are 1, or zero on a log scale. With a Wald test:

```
. xi: logistic lowbw i.agegrp
. testparm _Iagegrp_30 _Iagegrp_35 _Iagegrp_40
 ( 1)  _Iagegrp_30 = 0
 ( 2)  _Iagegrp_35 = 0
 ( 3)  _Iagegrp_40 = 0

        chi2(  3) =     1.58
    Prob > chi2 =     0.6649
```

The `testparm` line can be abbreviated to `_I*`.

14.5   Other regression commands

Both the likelihood ratio test and the Wald test can be used with any of the other
regression commands. With `regress`, however, they give essentially the same result
because the appropriate log likelihood is Gaussian and the quadratic approximation
to this log likelihood is perfect. There is a very small difference in the output, as you
will see in the following example. First the LR test:

```
. regress bweight hyp
. estimates store tmp
. regress bweight
. lrtest tmp

likelihood-ratio test          LR chi2(1)   =     29.02
(Assumption: . nested in tmp)  Prob > chi2 =    0.0000
```

then the Wald test:

```
. regress bweight hyp
. testparm hyp
 ( 1)  hyp = 0

F(  1,   498) =    29.76
     Prob > F =     0.0000
```

The difference is that the LR test uses the chi-squared distribution to obtain the p-
value, while the Wald test with `regress` uses the F distribution. The use of F tries
to take account of the fact that the residual variability in the response is estimated
from the data. The two tests give almost the same answer provided the dataset is not
too small.

Exercises

1. Load the diet data and find the effect of `hieng` as a rate ratio using Poisson regression. Test whether this effect is significantly different from 1, using both a LR test and a Wald test.

2. Test whether the effect of `energy` per kcal is significantly different from 1, using both a LR test and a Wald test.

3. Load the births data and find the effect of `hyp` on `lowbw` as an odds ratio, using `logit`. Test whether this effect is significantly different from 1 using both a LR test and a Wald test.

Chapter 15

Controlling and stratifying with regression

This chapter shows how to use the regression commands to find the effects of one variable controlled for another, how to find the effects of one variable stratified by another, and how to test for effect modification using interactions.

15.1   Controlling with regression commands

Regression commands can have more than one explanatory variable. As an example we shall use the births data to study how `bweight` varies with both `gestwks`, which is metric, and `sex`, which although categorical, takes only two values and can therefore be treated as metric. Try

```
. use births, clear
. regress bweight gestwks sex
```

The interpretation of the coefficients for `gestwks` and `sex` is as follows:

- The effect of a unit change in `gestwks` when `sex` is kept constant (i.e. when we control for `sex`) is an increase of 196 g of birth weight.

- The effect of a unit change in `sex` (from male to female) when `gestwks` is kept constant (i.e. when we control for `gestwks`) is a decrease of 189.9 g in birth weight.

We can also find the effect of `gestwks` controlled for `sex` using `effects`. Try

```
. effects, res(bweight) typ(metric) exp(gestwks) exm catcon(sex) md
```

The answer should be 196 g. Similarly, the effect of `sex` controlled for `gestwks` could be found by declaring `sex` as a categorical exposure and `gestwks` as a metric control variable. Try

```
. effects, res(bweight) typ(metric) exp(sex) exc metcon(gestwks) md
```

and you should get −189.9 g.

The `effects` command treats one variable as the exposure and the other as control and reports the effect of the exposure variable controlled for the control variable. Regression commands treat the two variables symmetrically, and report the effects of each variable controlled for the other. As another example we shall use `lowbw` as the response, and look at the effects of `gestwks` controlled for `sex` and the effects of `sex` controlled for `gestwks`. When effects are measured using odds ratios the appropriate command is

```
. logit lowbw gestwks sex, or
```

The effect of a unit increase in `gestwks`, controlling for `sex`, is to reduce the odds of low birth weight by a factor of 0.405, and the effect of a unit increase in `sex` (i.e. from male to female) controlled for `gestwks` is to increase the odds of low birth weight by a factor of 1.510.

15.2 Testing effects after controlling

This can be done using either the LR test or the Wald test. For example, to test the effect of `gestwks` on `lowbw` controlled for `sex` using the LR test, try

```
. logit lowbw gestwks sex
. estimates store tmp
. logit lowbw sex if e(sample)
. lrtest tmp
```

The first `logit` is based on 490 subjects because there are 10 missing values for `gestwks`. Without the `if e(sample)` the second `logit`, which does not use `gestwks`, would be based on 500 subjects, rendering the test invalid. Including the `if e(sample)` with the second `logit` makes sure it is based on the same subjects as the first. Note that it is the variable you want to test that is left out when fitting the second model - this is because you want to find out how much the log likelihood goes down when the parameter for that variable is set to its null value. To do the same thing with the Wald test, try

```
. logit lowbw gestwks sex
. testparm gestwks
```

15.3 Stratifying with regression commands

The effect of gestwks on bweight stratified by sex is easily found with

. effects, res(bweight) typ(metric) exp(gestwks) exm str(sex) md

This is more difficult to do with regression commands. The command

. regress bweight gestwks

shows the effects of gestwks on bweight and the command

. regress bweight gestwks sex

shows the effect of gestwks on bweight controlled for sex, but what we want to see in a stratified analysis is the effect of gestwks on bweight separately for each sex. For this we need to generate indicators for the two levels of sex and then multiply them by gestwks as follows:

. tabulate sex, gen(sex)
. gen gw_sex1 = gestwks*sex1
. gen gw_sex2 = gestwks*sex2

The variable gw_sex1 is the same as gestwks for male babies but zero for female babies and gw_sex1 is the same as gestwks for female babies but zero for male babies. The command

. regress bweight sex2 gw_sex1 gw_sex2

will produce the stratified effects opposite the variables gw_sex1 and gw_sex2.

As another example we shall return to the diet data and look again at the effect of hieng stratified by job. To see what the result should be, try

. use diet, clear
. effects, res(chd) typ(failure) exp(hieng) exc str(job) rr fup(y)

There are three effects of hieng, one for each level of job. To do the same thing with regression commands we have to create indicators for both hieng and job and multiply them together as follows:

. tabulate hieng, gen(hieng)
. tabulate job, gen(job)
. gen he_job1=hieng2*job1
. gen he_job2=hieng2*job2
. gen he_job3=hieng2*job3

Note that hieng1, the first of the two indicators for hieng has been omitted, making level 1 the baseline for the effect of hieng. The command

```
. poisson chd job2 job3 he_job1 he_job2 he_job3, e(y) irr
```

will produce the stratified effects. As you see Stata offers no simple way of doing stratified analysis with regression commands. Until it does, we recommend that you use `effects`, or the Mantel-Haenszel commands described in Chapter 16.

15.4 Testing for effect modification and interactions

When the effects of an exposure vary between strata there is said to be effect modification. To see how to test for effect modification using regression commands we shall use the diet dataset and look at whether the effect of `hieng` is modified by `job`. The output from `effects` is:

```
Level of job   Effect   95% Confidence Interval
driver          0.410   [ 0.124 , 1.362 ]
conductor       0.655   [ 0.227 , 1.888 ]
bank            0.518   [ 0.212 , 1.267 ]
```

and we can investigate whether there is any effect modification using *interactions*. Because the three effects are rate ratios a natural way of comparing them is to take the ratios of the second and third against the first:

Effects of hieng level 2/1		
job	Effect	Ratio
1	0.410	0.410/0.410 = 1
2	0.655	0.655/0.410 = 1.597
3	0.518	0.518/0.410 = 1.262

The two ratios 1.597 and 1.262 are called *interactions* between the effects of `hieng` and the variable `job`. To produce the interactions using `poisson`, try

```
. xi: poisson chd i.hieng*i.job, e(y) irr
```

The * between `i.hieng` and `i.job` tells Stata that interactions between these two variables are required. You should see the following (abbreviated) table:

```
        chd |       IRR
------------+------------
  _Ihieng_1 |   .4102648
    _Ijob_2 |   1.136857
    _Ijob_3 |    .813427
_IhieXjob_~2 |   1.596755
_IhieXjob_~3 |   1.261973
```

The fourth and fifth rows show the two interactions, identified as _IhieXjob_~2 and _IhieXjob_~3. Although Stata goes to a lot of trouble to name the interactions appropriately, the only important thing to note is that the X in the names shows they are interactions. The first three rows of the table refer to the effect of hieng at the first level of job, and the effects of job at the first level of hieng. These three effects are of little interest – only the interactions are useful.

When the three effects of hieng at the different levels of job are the same (i.e. there is no effect modification) the interactions will both be 1, apart from random variation. The test for no effect modification is therefore the same as the test that the interaction parameters are 1 (or 0 on a log scale). Try

```
. testparm _IhieXjob*
```

to test for interactions using a Wald test. You will see that the chi-squared statistic on 2 df is 0.33, and the p-value is 0.8475, so there is no evidence of effect modification.

To do the the same thing with a likelihood ratio test, try

```
. xi: poisson chd i.hieng*i.job, e(y) irr
. estimates store tmp
. xi: poisson chd i.hieng i.job, e(y) irr
. lrtest tmp
```

The second poisson command fits the model without the interaction terms. In this case the interactions between hieng and job do not differ significantly from 1, so we use the output from the second model (without interactions) to estimates the effect of hieng controlled for job.

```
. xi: poisson chd i.hieng i.job, e(y) irr
```

chd	IRR
_Ihieng_2	.5247666
_Ijob_2	1.358442
_Ijob_3	.8843023

The effect of hieng controlled for job is 0.525. The other two terms are the effects of job controlled for hieng.

15.5   Interactions with metric variables

If you ask for interactions when one variable is categorical and the other is metric the metric variable must be typed after the *. If you do it the other way round, like this

```
. xi: poisson chd height*i.hieng, e(y) irr
```

you will get an error message. But

. xi: poisson chd i.hieng*height, e(y) irr

will give the following (abbreviated) table:

```
        chd |        IRR
------------+------------
  _Ihieng_1 |   99.00107
     height |   .9258995
_IhieXheig~1 |   .9706023
```

As before the effects in the first two rows are not useful – only the interaction is of interest.

When both variables are metric we cannot use xi to include interaction terms. Instead a new variable needs to be created before running the regression. Try:

. gen energy100=energy/100
. gen enXht = energy100*height
. poisson chd energy100 height enXht, e(y) irr

The third row shows the interaction. The significance of this interaction is tested with

. testparm enXht

Exercises

1. Load the births dataset and use logistic to find the effect of hyp on preterm as an odds ratio.

2. Cut matage into 2 groups with

   . egen mage2=cut(matage), at(20,40,45) label

   and use logistic with interactions to study whether mage2 modifies the effect of hyp on preterm. Test the interaction with testparm and lrtest.

3. Check your answer using effects.

4. Use regress to find the effect of gestwks on bweight as a change in bweight per week of gestation.

5. Study whether this effect is modified by sex using testparm (remember that the categorical variable needs to be first when using *).

6. Check your answers using effects.

Chapter 16

Mantel-Haenszel methods

Mantel-Haenszel methods[6] were introduced in epidemiology as a way of controlling for potentially confounding variables in case-control studies by first creating strata based on the variables. Stratum-specific estimates of the effect of exposure, measured as odds ratios, are then obtained. Provided these are not too different, they are combined in a simple fashion, to provide a single estimate of exposure controlled for strata. The result is only an approximation to the fully efficient estimate provided by logistic regression, but because the method could be carried out by hand-calculator at a time when computers were not widely available, it became (and stayed) very popular. One particular benefit is that it can be used even when the individual strata contain very little data, as is the case with individually matched case-control studies.

16.1 The method

For an exposure with two levels the results for the any given stratum can be summarized as an odds of D_1/H_1 for the exposed group in that stratum, and D_0/H_0 for the unexposed group in that stratum. Here D_1, H_1 refer to exposed cases (D for Deaths) and exposed controls (H for Healthy), respectively. Similarly D_0, H_0 refer to unexposed cases and controls. The odds ratio which measures the effect of exposure in the stratum is

$$\frac{D_1/H_1}{D_0/H_0} = \frac{D_1 H_0}{D_0 H_1}$$

In the Mantel–Haenszel method, these stratum-specific odds ratios are combined as

$$\frac{\sum D_1 H_0/T}{\sum D_0 H_1/T}$$

where summation is over strata, and $T = D_1 + H_1 + D_0 + H_0$, the total number of subjects in the stratum.

16.2 The Stata command

As an example we shall use the births data and look at the effect of `hyp` on `lowbw`, measured as an odds ratio and stratified by `sex`. Although this was not a case-control study, the effect of exposure was measured as an odds ratio, so the Mantel-Haenszel method can be applied. To see what the result should be, using `effects`, try

```
. use births, clear
. effects, res(lowbw) typ(binary) exp(hyp) exc str(sex) or
```

```
Level of sex     Effect
1                5.316
2                2.773
```

The effect of `hyp` as an odds ratio is 5.316 for the first stratum, and 2.773 for the second. To combine these and obtain the effect of `hyp` controlled for `sex`, try

```
. effects, res(lowbw) typ(binary) exp(hyp) exc catcon(sex) or
```

```
Levels      Effect
2/1         3.906
```

To do this with Mantel-Haenszel, try

```
. mhodds lowbw hyp, by(sex)
```

```
sex | Odds Ratio
----+-----------
  1 |   5.316129
  2 |   2.773333
```

```
Mantel-Haenszel estimate controlling for sex
```

```
 Odds Ratio
 -----------
   3.896980
```

In the `mhodds` command the first variable is the response (`lowbw`), the second is the exposure (`hyp`), and the stratifying variable (`sex`) is in the `by( )`. The stratum specific odds ratios should be the same for the two methods, but the result of combining them using `mhodds` differs from the result using `effects` because the Mantel-Haenszel method is only an approximation to the regression methods on which `effects` is based. In fact the approximation is very good unless the stratum specific odds ratios are much larger or much less than 1.

To suppress the stratified effects (useful when there are lots of strata) put the stratifying variable after the exposure variable, as in

. mhodds lowbw hyp sex

To control for two variables it is necessary to create strata corresponding to all combinations of values of the two variables. For example, to control the effect of hyp on lowbw for both sex and preterm, avoiding missing values for preterm, try

. mhodds lowbw hyp if preterm < . , by(sex preterm)

Note that there are 4 strata corresponding to the 4 combinations of the 2 levels for sex and the 2 levels for preterm. To see the effects of hyp controlled for sex stratified by preterm, try

. mhodds lowbw hyp sex if preterm < . , by(preterm)

16.3   Matched case-control studies

In matched case-control studies controls are deliberately matched to the cases on variables which are likely to confound the effect of exposure. Such studies present two problems to epidemiologists. The first is that, contrary to intuition, it is still necessary to control for the matching variables, even though, as a result of matching, they are not associated with the disease outcome (case/control). Failure to do this results in a biased estimate of the effect of exposure. This is a consequence of using odds, and cannot be avoided with case-control studies. To illustrate the point, consider the made-up data in Table 16.1. There are 100 cases and 100 controls in each stratum, so the disease outcome is independent of strata. The odds ratio in each stratum is close to 2.0, so intuitively, because the disease outcome is independent of strata, we would expect the odds ratio when strata are ignored also to be 2.0. The fact that it is 1.7 is entirely due to the fact that we are using odds ratios.

Table 16.1: Bias due to ignoring matching

| Stratum | Cases | | Controls | | Odds |
	Exposed	Unexposed	Exposed	Unexposed	ratio
1	89	11	80	20	2.02
2	67	33	50	50	2.03
3	33	67	20	80	1.97
Total	189	111	150	150	1.70

The second problem arises as a consequence of the first. When there are many small matched sets, as there are with individual matching, there are too many strata to use logistic regression to control for the matching. As an example we shall use the Salmonella data, first referred to in Chapter 13, in which there are 59 matched sets each containing one case and either one or two controls. We shall attempt to find the effect of `plant7` after controlling for the matching using logistic regression, but first we need to increase the number of variables that Stata can include in an estimation command with `set matsize`. Try

```
. use salmonella, clear
. set matsize 100
. xi: logistic case plant7 i.set
```

You will first see the effect of `plant7`, which is 10.48, and then a lot of effects of `set`. Because of the large number of parameters and the small amount of data in each stratum, we cannot rely on this result.

Conditioning on the total number of exposed subjects in each stratum gives a likelihood which depends only on the common odds ratio, so using this conditional likelihood avoids the problem of too many parameters. The `clogit` command, introduced in Chapter 13, uses the conditional likelihood:

```
. clogit case plant7, group(set) or
```

Now the effect of `plant7` is 4.47, very different from the 10.48 obtained with logistic regression. The Mantel-Haenszel method of controlling is in fact an approximation to the conditional likelihood used in `clogit` not the likelihood used in `logistic`. For this reason it can be used to analyse matched case-control studies in which there are many strata and not much data for each stratum. For example,

```
. mhodds case plant7 set
```

gives 4.58, very close to the value obtained with `clogit`. It is best to suppress the stratum specific estimates by putting `set` after the exposure, not in `by( )`, because there are rather a lot of them.

16.4 Mantel-Haenszel methods for rates

The Mantel-Haenszel method can also be used in other types of study. To illustrate its use in follow-up studies we shall return to the diet data and look at the effect of `hieng` stratified by `job`. To see the result using `effects`, try

```
. use diet, clear
. effects, res(chd) typ(failure) exp(hieng) exc str(job) rr fup(y)
```

You should see the following (abbreviated) table:

```
Level of job     Effect

driver           0.4103
conductor        0.6551
bank             0.5177
```

The effect of `hieng` controlled for `job` is obtained with

```
. effects, res(chd) typ(failure) exp(hieng) exc catcon(job) rr fup(y)
```

To do this with Mantel-Haenszel[1], try

```
. mhrate chd hieng, by(job) e(y)
```

The first variable following the command is the response (`chd`), the second is the exposure (`hieng`) and the stratifying variable is in the `by( )`. In addition, for `mhrate` the follow-up time must be given in `e( )`.

16.5   Exposures on more than two levels

Mantel-Haenszel methods can be used to deal with exposures on (say) three levels, by comparing level 2 with level 1 and then level 3 with level 1. The Stata command must be run twice to do this. For example, try

```
. egen eng3=cut(energy), at(1500,2500,3000,4500)
. mhrate chd eng3, compare(2500,1500) e(y)
. mhrate chd eng3, compare(3000,1500) e(y)
```

Note that the actual codes for the levels are used inside the `compare`. If the `compare` is omitted the exposure is assumed to be metric. For example,

```
. generate energy100=energy/100
. mhrate chd energy100, e(y)
```

shows that an increase of 100 units of energy reduces the rate ratio by a factor of 0.901. This is close to the value obtained from

```
. poisson chd energy100, e(y) irr
```

which is 0.891.

---

[1]The command `mhrate` is not part of official Stata, but is included with the files which come with this book. The official Stata version of this command is `stmh` – see Chapter 17.

Exercises

1. Load the births dataset and use `mhodds` to find the effect of `hyp` on `preterm` as an odds ratio.

2. Cut `matage` into 2 groups with

   ```
   . egen mage2=cut(matage), at(20,40,45) label
   ```

   and use `mhodds` to study whether `mage2` modifies the effect of `hyp` on `preterm`.

3. Check your answer using `effects`.

Chapter 17

Survival data and stset

In this chapter you will learn how to declare data as `st`; how to summarize `st` data; how to use exponential regression to calculate rate ratios; how to split the follow-up time to take account of rates that vary during follow-up; and how to use Cox regression to do the same on a continuous time scale.

17.1   The response in survival data

The response in survival time data consists of two pieces of information – the time which the subject spends in the study, and what happens at the end of the study, usually held as a failure indicator. Stata has a family of commands, with names starting with `st` (where `st` stands for survival time), that deals with this type of data and for which the time and failure variables are specified once and for all using the command `stset` (survival time set). Once `stset` has been used the data are said to be in `st` form. Only then can the other `st` commands be used.

We shall illustrate this with a dataset called `pbc.dta`, which holds the data for 184 subjects who enrolled in a randomized clinical trial for the treatment of primary biliary cirrhosis, a chronic, but eventually fatal disease of the liver[1]. Load the PBC data and examine them with

```
. use pbc, clear
. describe
. tabulate d
```

Subjects were randomly allocated to two treatments, active and placebo, identified by the variable `treat`, coded 1 for placebo, 2 for active. The time variable (in years) is y and the failure variable is `d`, coded 1 for death from any cause, 0 otherwise. To declare the survival variables use

```
. stset y, fail(d)
```

The Stata output reports an overall summary. In this case there are 96 failures during a total follow–up time of 747 years so the overall rate is $96/747 = 0.129$ failures per year.

The failure variable is usually coded 1 for failure and 0 otherwise, but when different kinds of failure are recorded in the same study they can be coded with different numbers, e.g. the International Cause of Death code (ICD). To study all failures you would use `fail(d)` in the `stset` command, but to study particular kinds of failure, e.g. when `d` is coded 198 or 199, you would use `fail(d=198,199)`.

17.2   Summarizing survival time

The distribution of survival time cannot be observed directly, because of the subjects who have not failed by the end of the study. Instead we use the Kaplan-Meier survivor function, obtained with

```
. sts list
```

The first 5 columns of the first few lines of output look like this:

Time	Beg. Total	Fail	Net Lost	Survivor Function
.0219	184	0	1	1.0000
.0246	183	2	0	0.9891
.052	181	1	0	0.9836
.104	180	2	0	0.9727

where time is in years, so the earliest recorded time in the data is 0.0219 years, or 8 days. In the same row we read that 184 subjects were in the study just before 0.0219 years, none failed at that time, but one subject was lost (censored). Thereafter two fail, then one fails, then two fail, and so on.

The survivor function is the predicted probability that a new subject will survive a given time. To see how it is calculated, imagine that the time scale is divided into many tiny instants of time (e.g. hours), starting at time 0. The probabilities of failure and survival, referred to as P(F) and P(S) respectively, at each of the four instants represented in the table are

Time	Beg. Total	Fail	Net Lost	P(F)	P(S)
.0219	184	0	1	0/184 = 0	1 − 0 = 1
.0246	183	2	0	2/183 = 0.0109	1 − 0.0109 = 0.9891
.052	181	1	0	1/181 = 0.0055	1 − 0.0055 = 0.9945
.104	180	2	0	2/180 = 0.0111	1 − 0.0111 = 0.9889

The survival function for a given time is calculated by multiplying P(S) for all successive instants up to that time.

```
Time                          Survivor function
------------------------------------------------
.0219                                 1 = 1
.0246                    1 x 0.9891 = 0.9891
.052             1 x 0.9891 x 0.9945 = 0.9837
.104   1 x 0.9891 x 0.9945 x 0.9889 = 0.9727
```

At first sight we seem to have forgotten the instants between the ones shown in the table, but because there are no deaths in these instants the survival probabilities corresponding to these instants are 1, and the survival function is not affected – it only changes its value at times when there are failures. For the same reason the survivor function does not change when there is a censoring event.

The actual values of the survivor function are useful when trying to understand the method, but in practice the graph of the survivor function is more useful. This is obtained with

```
. sts graph
```

which plots the survivor function against time. The flat stretches of the the plot correspond to periods of time during which there were no failures, and the drops occur at the times of the failures. From the graph, the predicted probability of surviving 2.6 years is 0.75, the predicted probability of surviving 5.4 years is 0.5, and the predicted probability of surviving 9.3 years years is 0.25. Another way of putting this is that a new subject will have a 25% chance of failing by 2.6 years, a 50% chance of failing by 5.4 years, and a 75% chance of failing by 9.3 years, and this is the summary produced by

```
. stsum
```

Note that it is not necessary to specify the survival variables with these commands because they have been specified with stset. Now try

```
. sts graph, by(treat)
```

to show the survival curves for the two treatment groups on the same graph, and

```
. stsum, by(treat)
```

to see the summary by treatment group. The median survival time is 5.3 years in the placebo group and 6.0 years, i.e. better, in the active treatment group.

17.3   Calculating rates and rate ratios

An alternative way of summarizing survival data is to calculate the rate at which failures are occurring, i.e. the number of failures per unit time. Try

. strate

which reports an overall rate of 0.129 deaths per year. To find the yearly rate for each treatment group, try

. strate treat

The rate is higher for the placebo group. The units of the rate are taken from the units of the time variable declared in stset, i.e. y (coded in years), but they can be changed. To calculate rates per 10 years, try

. strate treat, per(10)

To calculate the rate ratio for active versus placebo treatment, together with a confidence interval, the command streg can be used. This command allows you to fit regression models to survival times. Because the distribution of survival time is generally skewed, streg fits regression models with distributions like exponential and lognormal. When the rate is constant over the entire follow-up time, the distribution of the survival time is exponential, so the appropriate form of the command on this assumption is

. streg treat, dist(exp)

Here the option dist stands for distribution and the argument exp stands for exponential. The estimated rate ratio is 0.87, which is simply the ratio of the two yearly rates, 0.120 and 0.138. Note that the output refers to the Hazard Ratio, where the hazard is an alternative word for rate. We would have obtained the same result using the poisson command introduced in Chapter 13. To see this, try

. poisson d treat, e(y) irr

Because poisson is not an st command it requires the failure variable d as response, and the follow-up time y.

17.4 Variables created by stset

When stset is used, four new variables are created for internal use: it is not necessary for the user to know about these variables, but it helps when trying to understand what is going on. The variables are:

    _t0   time at entry
    _t   time at exit
    _d   failure indicator
    _st   inclusion indicator

By default the time at entry is 0 so the time at exit is equal to the follow-up time. The failure indicator takes the same values as the failure variable declared in stset.

The inclusion indicator, _st, flags all subjects to be included in any following st commands. Try

```
browse id _t0 y _t d _d _st
```

to see the values these variables take for the PBC data. You will see that _t0 is always zero, _t is the same as y, and _d is the same as d. The variable _st is 1 for all observations, which means they are all included. When an observation is excluded from the analysis for some reason, the value of _st is set to 0.

17.5   Rates that vary with time

In many applications the rates are not constant over the entire time in the study. One way of dealing with this is to split time into bands short enough to assume that during each band of follow-up the rate is constant (although it may change from one band to the next), and to estimate rates separately for each band. To split the times of the PBC data we need the command stsplit, but first we need to stset the data again so that the option id is included:

```
. stset y, fail(d) id(id)
```

The option id( ) declares the name of the variable which holds the subjects' identi fiers, also called id in this dataset, so that when the individual times are split they can still be linked back to the original subject. In order to see what happens we list the information for subject 45 before the split:

```
. list id  _t0 _t _d if id==45, noobs
```

```
 id   _t0            _t   _d
-------------------------------
 45    0    5.3114305    1
```

Then we split the follow-up into (say) 2-year bands with a 4-year band at the end (from 8 to 12 years), and list again with:

```
. stsplit timeband, at(0,2,4,6,8,12) trim
. list id _t0 _t _d timeband if id==45,noobs
```

```
 id   _t0            _t   _d   timeband
---------------------------------------
 45    0             2    0          0
 45    2             4    0          2
 45    4    5.3114305     1          4
```

Subject 45 spent 5.3 years in the study and this time has been split into 2 years in the band 0–2, at the end of which the subject was still alive (_d=0), 2 years in the band 2–4, at the end of which the subject was still alive (_d=0), and a last period of 1.3 years (from 4 to 5.3) in the band 4–6 at the end of which the subject failed (_d=1). The points at which time is split are defined by at(...), and the option trim makes sure that no follow-up time outside the range specified is included. The word timeband following stsplit declares the name to be given to the new variable that identifies the bands. This takes the value of the lower end of each band. Try

. tabulate timeband

which shows the values taken by timeband. It also shows that, although to begin with there were 184 subjects, after splitting their times there are 466 observations, i.e. around 2.5 per subject. The overall information on the follow-up times and failures of these subjects has not been corrupted however: it has only changed format. Try

. tabulate _d

and you will see that there are still only 96 failures.

The reason for using stsplit was to find the timeband-specific yearly rates, which you can do with

. strate timeband

These rates can be also plotted by adding the option graph:

. strate timeband, graph

We can estimate the rate ratio for active treatment versus placebo, controlled for timeband, with

. xi: streg treat i.timeband, dist(exp)

The rate ratio for active versus placebo treatment controlled for timeband is 0.84, which is not very different from the original overall rate ratio of 0.87.

To look at the effect of treatment stratified by timeband, using Mantel-Haenszel methods we can use the st version of mhrate which is stmh. Try

. stmh treat, by(timeband)

timeband	RR	Lower	Upper
0	1.15	0.60	2.20
2	1.03	0.45	2.37
4	0.54	0.22	1.31
6	0.52	0.18	1.56
8	0.55	0.05	6.04

The treatment appears to have no effect up to 4 years, with a beneficial effect after this. However, the test for effect modification (p=0.56) provides no evidence for such time-changing effects, so we should assume a constant effect of treatment and control for timeband using the second part of the stmh results which give

Overall estimate controlling for timeband

```
    RR    chi2   P>chi2   [95% Conf. Interval]
-----------------------------------------------
  0.841   0.73   0.3939     0.565      1.252
```

17.6 Cox regression

Sometimes rates vary so quickly with time that it would be necessary to split the follow-up into many pieces. In these situations, if we are only interested in estimating the rate ratio between different groups controlled for time, we can use `stcox` to fit what is generally called Cox regression, or the proportional hazards model. Proportional hazards imply that the hazard ratio (or rate ratio) for the groups being compared is the same at all times during follow-up. It is not necessary to split the data when using `stcox` because the splitting is implicit in this command, so try

```
. use pbc, clear
. stset y, fail(d)
. stcox treat
```

The effect of treatment controlled for time in the study is 0.856 which is close to 0.839, the value obtained using `streg` with `treat` and `i.timeband`. Using smaller and smaller time-bands in `stsplit`, the result from `streg` gets closer and closer to the result from `stcox`.

The main difference between these two approaches is that with Cox regression we can ignore how rates change with time, and they can vary freely, while with `streg` rates are assumed to be constant, or to vary in steps. However, both regression methods assume that the rate ratio between the two groups is constant over the follow-up time, and both report the effect of `treat` controlled for time; `stcox` controls for time continuously, while `streg` controls for time in bands.

Checking on the proportional hazards assumption is straightforward with `stmh` because the rate ratio is estimated for each timeband, and it is only a question of looking at these to see whether they are approximately the same. With `stcox`, when estimating the hazard ratio for two groups, it is possible to obtain a smoothed plot of the log of the underlying hazard against time, for each group, with

```
. sts, by(treat) hazard ylog
```

The two curves should be parallel if the proportional hazards assumption is true. In this case the hazards seem very close up to about 4 years, with a hazard ratio just greater than 1, and then proportional (parallel on a log scale) from 4 years onwards with a hazards ratio less than 1. In other words the effect of treatment seems to be modified by time in study. To test whether there really is effect modification, the simple approach is to split the data and use the methods described in Section 17.5.

Alternatively, to test whether there is effect modification using continuous time, you can use the Schoenfeld test. This is based on the scaled Schoenfeld residuals generated after a Cox model is fitted. If there are several explanatory variable in the model, there is a different set of residuals for each variable. If the residuals show a significant linear trend with time, there is evidence of lack of proportionality. Try

```
. stcox treat, schoenfeld(s)
. stphtest
```

where s is a new variable that holds the Schoenfeld residuals. For further reading about survival analysis we recommend *An Introduction to Survival Analysis for Stata, Revised Edition.*[3]

Exercises

1. Load the `cancer.dta` dataset. How many deaths occurred during the study?

2. Now `stset` the data and find the median survival time for each drug type. Plot the survival curves for the three groups.

3. Find the rates of death for each drug type, per year.

4. Find the rate ratios for each of the active drugs compared with placebo using `streg` with `xi`.

5. Break the follow-up time into 5 intervals, from 0 to 24 months in 6 month time-bands and then from 24 to 40 months. Call this variable `timeband`. Find the overall rates of death for each time-band, per year.

6. Use `streg` to find the effect of each active drug compared with placebo, controlled for `timeband`.

7. Use `stcox` to find the effect of each active drug compared with placebo.

Chapter 18

Different time scales and standardization

In clinical follow-up studies time is usually recorded as time since entry to the study, but in population studies time is more likely to be recorded as calendar time, i.e. as date of entry to the study and date of exit. Another difference is that in clinical follow-up studies the time since entry is usually the most important determinant of the rate of interest, whereas in population studies other time scales, such as age, may be more important. To deal with this `stset` has options to change the time scale used for analysis. In this chapter we explore these options.

18.1   Follow-up time

No matter how time is originally recorded it is always possible to calculate follow-up time and to use it as the time variable. For example, in the diet data, entry to and exit from the study are originally recorded as string dates and then converted to Stata format. The variable y was calculated from `(dox - doe)/365.25` and added to the dataset. This measures follow-up time in years, using 365.25 as the average number of days in a year. Using this variable we can proceed as in the previous chapter, and calculate rates and rate ratios with

```
. use diet, clear
. stset y, fail(chd)
. strate hieng, per(1000)
. streg hieng, dist(exp)
```

By starting in this way we are using the time since entry as the time scale, but it is better to `stset` using the original time variables `dox` and `doe` because then we can consider other time scales as well.

18.2 The diet data

In the diet data there are three variables which describe follow-up: doe which is the date of entry to the study, dox which is the date of exit, and chd which is the failure indicator. Remember that in Stata format dates are measured in days, with an arbitrary origin 0 at 1/1/1960. To stset the data using these variables, try

```
. stset dox, fail(chd) origin(time mdy(1,1,1900)) enter(doe)
      id(id) scale(365.25)
```

where we have included id(id) in case we need to split the data later. What stset does is to create new exit and entry times

```
_t  = (dox - origin)/scale
_t0 = (doe - origin)/scale
```

from dox and doe using the new origin and scale. Two points about the stset command are worth noting. The first is the origin of the original time scale is at 1/1/1960, but this has been shifted to mdy(1,1,1900), an operation which needed the keyword time. The reason for this is that some subjects entered the study before 1/1/1960, and stset ignores follow-up time before the origin. The second is that the option scale(365.25) converts days to years. To see the effect of this stset on the underlying variables, try

```
. list id _t0 _t if id==163
```

```
   id          _t0           _t
---------------------------
  163    60.788501    74.157426
```

Remembering that time is now measured in years from 1/1/1900, you see that subject 163 enters the study towards the end of 1960, and develops CHD at the beginning of 1974. The follow-up time is therefore $74.157 - 60.788 = 13.369$ years. To calculate rates per 1000 years, try

```
. strate, per(1000)
```

which shows that the overall rate is 9.99 per 1000 years. Similarly

```
. strate hieng, per(1000)
```

shows the rates for the two levels of hieng, and

```
. streg hieng, dist(exp)
```

shows the rate ratio. These results are the same as those obtained with

```
. poisson chd hieng, e(y) irr
```

18.3   Rates that change with time

In the above analysis the rates are overall rates, and we have made no effort to see whether they change with time, or to control rate ratios for time. Before doing this we should consider which time scale to use. The calendar time scale was that used to record the data, but calendar time may not be a major determinant of the rate. Age is likely to be more important. We can change the time scale from calendar time to age by re-defining the origin as the date of birth (held in `dob`):

```
. streset, origin(dob)
. list id _t0 _t _d if id==163, noobs
. strate, per(1000)
. strate hieng, per(1000)
```

Redefining the origin changes `_t0` and `_t` but does not change the overall rates because these depend only on the times between entry and exit, not on the origin.

To study how rates change with age we shall split the follow–up on the age scale using 5-year age bands. First check that the data are `stset` properly with

```
. stset
```

which should show that the origin is `dob` and that the time of entry is `doe`. If this is not the case, you will need to start again with

```
. stset dox, fail(chd) origin(dob) enter(doe) id(id) scale(365.25)
```

We now split the follow-up, ignoring the follow-up time before age 40 with the option `trim` (there were no failures before this age).[1] Before and after doing this we check on subject 163 to see the effects of the split:

```
. list id _t0 _t _d if id==163, noobs
. stsplit ageband, at(40(5)70) trim
. list id _t0 _t _d ageband if id==163, noobs
```

Before the split, the listing was

```
 id         _t0          _t   _d
------------------------------
163    47.55373   60.922656    1
```

This shows that subject 163 entered at age 47.55 and left at age 60.92 when he developed CHD. After the split, the listing is

---

[1]Any records referring to age before 40 will have `_st` set to 0, and will be excluded from any further `st` analyses.

```
id        _t0            _t    _d
------------------------------------
163    47.55373          50    0
163          50          55    0
163          55          60    0
163          60    60.922656    1
```

which shows that the single record for subject 163 has been replaced by four: the first covering the age band 45–49, the second covering 50–54, the third covering 55–59, and the fourth covering 60–64. The subject spends only 0.92 years in the final age band before developing CHD. Altogether 858 new observations have been created by this stsplit. To produce age–specific rates, try

. strate ageband, per(1000)

You can obtain a graph of the rates with

. strate ageband, per(1000) graph

Now try

. strate, per(1000)

and you will see that the overall rates have changed slightly from 9.99/1000, because some of the follow-up time has been trimmed by the stplit command.

To control the effect of hieng for ageband, try

. xi: streg i.hieng i.ageband, dist(exp)

This controls the effect of hieng for ageband by making two assumptions: one, that the rates are constant during each of the age-bands; two, that rate ratio between high and low energy groups does not change with ageband. To look at the effect of hieng stratified by ageband, try

. stmh hieng, by(ageband)

which also shows that the effect of hieng controlled for ageband, using the Mantel-Haenszel method, is 0.536, very close to that given by streg, which was 0.539.

18.4 Using non-st commands with st data

Great care must be exercised when using non-st commands with st data because st commands operate on the underlying time variables _t0 and _t, not on the original time variables doe and dox. The safest thing to do is to create new failure and follow-up variables for the underlying variables as follows:

```
. generate D = _d if _st==1
. generate Y = _t - _t0 if _st==1
. order D Y
```

The `if _st==1` is there in case any records were excluded, and the `order` command puts D and Y first in the list of variables, for convenience.

You can now use `db tabmore` to make a table of the age-specific rates by entering D as the *Response*, *failure* as the *Type*, Y as the *Follow-up time*, and *Rates* as the *Summary*. Similarly you can use `db effects` to find the effect of `hieng` controlled for `ageband`. You should get the same results as those just obtained with `strate` and `streg`. The command

```
. xi: poisson D i.hieng i.ageband, e(Y) irr
```

will give the effect of `hieng` controlled for `ageband`, and

```
. mhrate D hieng, by(ageband) e(Y)
```

will give the Mantel-Haenszel stratified analysis.

18.5   Two time-scales

To see how rates are changing jointly with two time scales we need to split the follow-up time on both scales. To split the diet data on both age and calendar period, first split on age with

```
. use diet, clear
. stset dox, fail(chd) origin(dob) entry(doe) scale(365.25) id(id)
. stsplit ageband, at(40(5)70) trim
```

There are three alternative syntaxes for `stsplit`, and to re-set the origin, having already split the follow-up time, the second is required.[2] To split the data on calendar period, again using 5-year bands, try

```
. stsplit period, after(time=mdy(1,1,1900)) at(50(5)80) trim
. replace period = period + 1900
```

After this last split the variable `period` takes the value 50 for the calendar period 1950-54, 55 for 1955-59, etc., so for clarity, we have added 1900 to these values. After inspecting the results with

```
. list id _t0 _t _d ageband period if id==163, noobs
```

you will see there are now 6 records for subject 163. To calculate rates by age, try

---

[2]You cannot reset the origin and then split because this will also reset _st to 1, so follow-up which was trimmed in the first split will be included in the second.

```
. strate ageband, per(1000)
```

To calculate rates by calendar period, try

```
. strate period, per(1000)
```

To find the effect of `hieng` controlled for both `ageband` and `period`, try

```
. xi: streg i.hieng i.ageband i.period, dist(exp)
```

The answer is 0.552.

The command `strate` can only produce one-way tables of rates, so to produce two-way tables we need to use `tabmore`, which is a non-`st` command. First generate new failure and time variables with

```
. generate D = _d if _st==1
. generate Y = _t - _t0 if _st==1
```

Then produce tables of rates per 1000, by age and period, with

```
. tabmore, res(D) typ(failure) row(ageband) col(period) rate fup(Y)
```

18.6 Standardization

Mortality and incidence rates for many diseases vary strongly with age, and it is always necessary, when comparing rates between different groups of subjects, to control such comparisons for age. In the early days of epidemiology this was done with direct and indirect standardization, techniques still widely used today.

Direct standardization

In direct standardization the age–specific rates found in each of the groups being compared are applied to a population with a standard age distribution in order to predict the number of deaths (mortality studies) or new cases (incidence studies) that would have occurred in that population. Two commonly used standard age distributions are those of Europe and the World. The total number of deaths expected in a standard population of size (say) 100 000, using the age–specific rates for a study group, is the directly standardized rate per 100 000 for that study group. We shall illustrate the process with an example.

Data for mortality in England and Wales County Boroughs and Rural Districts for 1936 are in the file `cbrd.dta`. The problem is to compare the mortality between these two places taking account of differences in age. First load these data and see how they are organized. Then find the total deaths and total population, by place, and hence the two crude death rates per 1000, with

```
. use cbrd, clear
. collapse (sum) deaths pop, by(place)
. gen crude_rate=deaths/pop*1000
. list
```

To take account of possible differences in age structure between County Boroughs and Rural Districts we shall standardize the rates for age using the world population, which is in the file stndpop, To describe this file on disk and make sure it is sorted by agegrp, try

```
. describe using stndpop
```

Now load the cbrd data and create a rate variable (per 1000) with

```
. use cbrd, clear
. generate rate = deaths/pop*1000
. sort agegrp
. list
```

Merge with the standard population in stndpop, matching on agegrp, with

```
. merge agegrp using stndpop
```

To directly standardize the rates try

```
. gen dst_rate=rate*wt
. collapse (sum) dst_rate, by(place)
. list
```

Indirect standardization

Indirect standardization is used when the study groups being compared are small, so that their age–specific rates are unreliable. In this case age–specific reference rates, usually based on the country from which the study groups were selected, are used in place of a standard age distribution. For each of the study groups being compared the standard rates are used to predict the number of cases expected in each age–group. The total expected cases for a group is compared with the total observed cases to give a standardized ratio (observed/expected). In the case of mortality this is called the standardized mortality ratio (SMR). An SMR of 1 means that the rates in the study group are the same as the reference rates, and an SMR of 2 means that the rates in the study group are twice the reference rates.

We return to the 1936 mortality data for County Boroughs and Rural Districts in England and Wales. To carry out indirect standardization we use the England and Wales standard rates for the period 1901–1910, which are in stndrate.dta. Load the standard rates from stndrate and check that the file is sorted by agegrp. Now load the data from cbrd and merge with the standard rates, matching on agegrp with

```
. use cbrd, clear
. sort agegrp
. merge agegrp using stndrate
```

Calculate the expected number of deaths from standard rates for each `agegrp` and `place` and put it in `exp`, with

```
. gen exp=rate*pop/1000
```

Then sum both observed and expected deaths over `agegrp`, and hence calculate the SMR for county boroughs and rural districts with

```
. collapse (sum) deaths exp, by(place)
. gen smr=death/exp
. list
```

You can do all this with the command `smrby`[3], as follows:

```
. use cbrd, clear
. sort agegrp
. merge agegrp using stndrate
. gen exp=rate*pop/1000
. smrby deaths exp, by(place)
```

---

[3]The command `smrby` is not part of official Stata but was contributed by Peter Sasieni. It is included with the files that come with the book.

Exercises

1. The file `mortality.dta` contains the deaths and population for Sweden and Panama, by age. Find the crude death rate per 100 000 for each country. The file `stndpop3.dta` contains the standard world population for the same age groups. Find the directly standardized rates for Sweden and Panama using this standard population.

2. The file `asthma.dta` contains data for the 403 counties of England and Wales which are to be compared in respect of deaths from asthma, recorded over a period of 5 years. Load the data and describe them. For each county and age–sex group the variable `pop` contains the number of subjects alive at the middle of the 5 year period and the variable `deaths` contains the number of deaths during the 5 year period. The file is sorted on `age` and `sex`. The total time spent in the study is approximately the mid–year population × 5 years, which is in the variable `pyrs`.

   (a) Find the crude rates per 100 000 for each county.

   (b) To obtain standardized mortality ratios for asthma, first have a look at the file `ewrate_asthma.dta`. The variable `rate` contains the reference rate (per 100 000), for each age–sex group, for the period of the study. Then merge `asthma.dta` with `ewrate_asthma.dta`, and calculate the expected deaths for each county. Use `collapse` to sum the observed and expected deaths over counties and calculate the SMR's. You will see a wide variation in SMR, from 0 to around 2.5.

Chapter 19

Writing Stata programs

The simplest sort of Stata program is a do file. However, although a do file might be perfectly adequate for personal use, it would probably not be useful to others. Some degree of generality is required before a program can be passed to others. In this chapter you will learn how to write Stata programs by starting with a do file, and gradually generalizing it so that it is useful to others.

19.1 Starting with a do file

The process will be illustrated with a simple do file which adds a straight line to a scatter plot. The do file is called `lgraph.do` and contains the commands

```
regress bweight gestwks
predict p
twoway (scatter bweight gestwks) (connected p gestwks, sort msymbol(i))
```

The first command fits a linear regression of `bweight` on `gestwks`, the second predicts the birth weight for each baby from its gestational age, and the last draws the scatter plot and joins up the predicted values to form a line. This do file is included in the same directory as your data files, so to run it all you need do is

```
. use births, clear
. do lgraph
```

Of course this can be done more efficiently in Stata 8 using the `lfit` command, but the example will nonetheless serve to demonstrate the basics of Stata programming.

19.2 Making the do file into an ado file

The file `lgraph1.ado` contains the following commands (*don't* type them):

```
*! program to add a line to a scatter plot, first try
program define lgraph1
version 8.2
regress bweight gestwks
predict p
twoway (scatter bweight gestwks) (connected p gestwks, sort msymbol(i))
end
```

The first line is a comment to remind you what the program does; the second states that the command `lgraph1` is defined by this program, which must be in a file called `lgraph1.ado`; the third states which version of Stata is being used, and the last line indicates the end of the program. Programs with the extension `.ado` are automatically loaded when called, so that in subsequent calls they are already in memory, and time is saved. To run `lgraph1.ado`, try

```
. use births, clear
. lgraph1
```

There is one serious thing wrong with this program which will be apparent if you run it again with

```
. lgraph1
```

Because the variable `p` already exists the program cannot create it, and Stata gives an error message. Of course you could start by dropping `p`, before running `lgraph1`, but there is a better way of dealing with the problem. This is to use a temporary variable for an intermediate result like `p`, which is automatically removed after the command has finished. The result of making this change is stored in the file `lgraph2.ado` which contains:

```
*! program to add a line to a scatter plot, second try
program define lgraph2
version 8.2
tempvar p
regress bweight gestwks
predict 'p'
twoway (scatter bweight gestwks) (connected 'p' gestwks, sort msymbol(i))
end
```

The line `tempvar p` defines a temporary variable name, generated by Stata, which is guaranteed not to clash with a name already in the dataset. This name, usually something like `__0000SD`, is stored in a local macro called `p`, and the contents of the macro are referred to as `'p'`. This means that you can refer to the temporary variable as `'p'`, and you don't have to know the actual name. A *macro* is best thought of as a box, with a name, which contains a string of characters. Try

```
. lgraph2
. lgraph2
```

Temporary variables are automatically dropped when a program finishes, so the program can be repeated a second time without clashing with variables created on the first run.

19.3 Cutting out unwanted output

All the output that comes with the `regress` command is not wanted here, because `regress` is used only to get the predicted values. The prefix `quietly` before a command prevents all output except error messages. Some such prefixes have been introduced and stored in the file `lgraph3.ado` which contains:

```
*! program to add a line to a scatter plot, third try
program define lgraph3
version 8.2
tempvar p
quietly regress bweight gestwks
quietly predict 'p'
twoway (scatter bweight gestwks) (connected 'p' gestwks, sort msymbol(i))
end
```

Try this now with

```
. lgraph3
```

and you will see that all the regression output has disappeared.

19.4 Making the program accept arguments

At present the command will only plot `bweight` against `gestwks`, which is not very useful as a general tool. It would be much more useful if the command allowed us to name the variables in the command line, so that `lgraph3 bweight matage` would add a line to the plot of `bweight` against `matage`, and so on. This is done by using the special command `args` (short for arguments). The file `lgraph4.ado` contains:

```
*! program to add a line to a scatter plot, fourth try
program define lgraph4
version 8.2
args y x
tempvar p
quietly regress 'y' 'x'
quietly predict 'p'
```

```
twoway (scatter 'y' 'x') (connected 'p' 'x', sort msymbol(i))
end
```

This would be run with commands like

```
. lgraph4 bweight gestwks
. lgraph4 bweight matage
. lgraph4 matage hyp
```

When the command `lgraph4 bweight gestwks` is executed, the line `args y x` in the program ensures that the local macro y contains the name of the first variable, bweight, and the local macro x contains the name of the second variable, gestwks.

Although this works, it is better to include a syntax line in the program, as follows:

```
*! program to add a line to a scatter plot, fifth try
program define lgraph5
version 8.2
syntax varlist(min=2 max=2)
tokenize 'varlist'
args y x
tempvar p
quietly regress 'y' 'x'
quietly predict 'p'
twoway (scatter 'y' 'x') (connected 'p' 'x', sort msymbol(i))
end
```

The `syntax` command states that `lgraph5` expects a list of two variable names (min and max both 2), and it places them in the local macro `varlist`. The `tokenize` command breaks `'varlist'` into the individual variable names and puts them in the local macros 1 and 2 (called tokens). The `args` command copies the contents of these tokens into local macros y and x. To see the advantage of `lgraph5` over `lgraph4`, try

```
. lgraph4 bweight gestwks,
```

which gives an incomprehensible error message, and

```
. lgraph5 bweight gestwks,
```

which gives an error message that options are not allowed. Options follow a comma, so typing a comma has alerted the command to options, but these are not allowed by the syntax.

19.5   Allowing if, in, and options

It is likely that the command `lgraph5.ado` is now sufficiently general to be a useful personal program, but to make it more widely available it should also respond to other aspects of the Stata command line like if and in. This is also done with the syntax command. The program lgraph6.ado contains:

```
*! program to add a line to a scatter plot, sixth try
program define lgraph6
version 8.2
syntax varlist(min=2 max=2) [if] [in] [, *]
tokenize 'varlist'
args y x
tempvar p
quietly regress 'y' 'x' 'if' 'in'
quietly predict 'p' 'if' 'in'
twoway (scatter 'y' 'x' 'if' 'in')
       (connected 'p' 'x', sort msymbol(i)), 'options'
end
```

The syntax command states that lgraph6 must be followed by a list of two variables, and that if and in are optional. The square brackets are used to denote the fact that they are optional. The * after the comma also in square brackets, indicates that any options can be included. The syntax command automatically places the list of variable names in the local macro varlist. The tokenize and args commands then copy the variable names into the local macros y and x. When you type the command

. lgraph6 bweight gestwks if hyp==1

for example, the local macros y, x, and if will contain bweight, gestwks and if hyp==1, respectively, so

quietly regress 'y' 'x' 'if' 'in'

will get translated into

quietly regress bweight gestwks if hyp==1

and so on throughout the program. Similarly, when you type

. lgraph6 bweight gestwks, xlabel(20 45) ylabel(1000 5000)

the local macro options will contain xlabel(20 45) ylabel(1000 5000). Of course the options you include must make sense in the command you attach them to, otherwise there will be an error message.

19.6 Discarding previous versions of a program

In the above we have shown a sequence of programs called lgraph1, lgraph2, etc., because this was a convenient way of showing what was going on. In practice you would use the *Do-file Editor* (or some other text editor) to upgrade your program step by step, without changing its name. The sequence would be something like this:

1. Use the *Do-file Editor* to create the original do file, called `lgraph.do`.

2. Check that it works with `do lgraph`.

3. Use the *Do-file Editor* to add the `program define`, `version` and `end` lines, and to save it as `lgraph.ado`.

4. Use the *Do-file Editor* to improve your program, checking at each stage that it works. It is not necessary to close the *Do-file Editor* during this process.

The fact that programs with the extension `.ado` are automatically loaded into memory when called, and kept there, is generally helpful because it speeds things up, but when writing programs it becomes a disadvantage. For example, after the first call to `lgraph` you might make a change to the file `lgraph.ado`, but if you now tried to run it you would be using the old `lgraph`, because it is still in memory. To get the new `lgraph` you need to issue the command

`. discard`

which drops all programs from memory. Failure to remember to do this can result in a very frustrating experience. You know you made the change, but it will not show up when you run the command! The reason is that you are running the previous version of the command, not the current one.

19.7   Another example: counting distinct values

In this example we shall write a command which finds the number of distinct combinations of values taken by a list of variables. First we must think of a name for the command, and we shall try `distinct`. You can check whether a name already exists by running it without any arguments:

`. distinct`

Stata complains that there is no such command, so it is OK to use this name. If the command exists, Stata will either execute it, or complain that there is something wrong with it.

Now we must create a do file which does the job in a special case, for example finding the number of distinct values taken by `sex` and `hyp` (the answer is 4, namely 1 0, 1 1, 2 0, 2 1). The basic idea is to sort on `sex` and `hyp`, and then to use `by` to count 1 for each new combination of values taken by `sex` and `hyp`. Put the following commands in a file called `distinct.do`, and run it with `do distinct`:

```
bysort sex hyp: generate u=1 if _n==1
tabulate u
display r(N)
```

The variable u takes the value 1 only for the first record in each combination, so when it is tabulated the number of 1's should be the number of distinct combinations. The command `tabulate` returns `r(N)` equal to the number of observations in the table.

We can make this into a program by adding the lines

```
*! program to find the number of distinct values
program define distinct, sortpreserve
version 8.2
```

at the beginning, and the line

```
end
```

at the end, and saving the result as `distinct.ado` in your working directory. Because the program changes the sort order of the data, which may have unforseen consequences, it is worth adding the option `sortpreserve` to the `program define` line to make sure the sort order is returned to whatever it was, when the program finishes. Run this command to make sure it works with

```
. drop u
. distinct
```

The first improvement is to replace the variable u with a temporary variable and to eliminate some unwanted output:

```
*! program to find the number of distinct values
program define distinct, sortpreserve
version 8.2
tempvar u
quietly bysort sex hyp: generate `u'=1 if _n==1
quietly tabulate `u'
display r(N)
end
```

Run the command to make sure it works with

```
. discard
. distinct
```

The next stage is to allow any variables, not just `sex` and `hyp`. The file should now contain

```
*! program to find the number of distinct values
program define distinct, sortpreserve
version 8.2
syntax varlist(min=1)
tempvar u
```

```
quietly bysort 'varlist' : generate 'u'=1 if _n==1
quietly tabulate 'u'
display r(N)
end
```

The syntax command asks for at least one variable, and puts the list of variables in the local macro called `varlist`. When you run this with

```
. discard
. distinct sex hyp
```

the local macro `varlist` will contain `sex hyp`, and when you run it with

```
. distinct matage
```

the local macro `varlist` will contain `matage`.

To demonstrate the use of a specific option we shall add an option to print the total number of records, as follows:

```
*! program to find the number of distinct values
program define distinct, sortpreserve
version 8.2
syntax varlist(min=1) [, Total]
tempvar u
quietly bysort 'varlist' : generate 'u'=1 if _n==1
quietly tabulate 'u'
display r(N)
if "'total'" == "total" {
  display _N
}
end
```

The addition to the syntax line states that the option `total` is allowed, and that it can be abbreviated to `t`. The capital letters in an option indicate the minimum abbreviation. The line

```
if "'total'" == "total" {
```

checks whether the local macro `total` contains the string `"total"`. Try this now with

```
. discard
. distinct hyp sex, total
. distinct hyp sex, t
```

Finally we shall improve the output by replacing the last five lines with

```
display as text "Number of distinct values is " as result r(N)
```

```
if "'total'" == "total" {
  display as text "Total number of record is " as result _N
}
end
```

where, on the standard Stata black background, `display as text` means display in green, while `display as result` means display in yellow. Try this with

```
. discard
. distinct hyp sex, total
. distinct matage, total
```

and check the last one with

```
. codebook matage
```

19.8   Some additional programming points

The `syntax` command puts the list of variable names in the local macro 'varlist'. To do something to each variable in turn, e.g. `summarize`, the following commands are used:

```
foreach var in local varlist {
    quietly summarize 'var'
    display "Total for variable 'var' is " r(total)
}
```

A simple way of allowing `if` and `in`, without including them in all the relevant program lines, is to use the command `preserve` as follows:

```
if "'if''in'" != "" {
  preserve
  keep 'if' 'in'
}
```

The first line checks that there is something in the `if` and `in` local macros, and if there is, the next line preserves the data as it is, and the last line keeps only those records satisfying the restrictions. Once the program has finished, the data are automatically restored to what was in memory just before the `preserve`. If necessary, the data can be restored before the end of the program with the command `restore`.

For a program to generate a new variable requires the syntax

```
. syntax ... , GENerate(string)
```

where `GEN` is the smallest acceptable abbreviation for `generate`. When the program is called with `gen(abc)` the local macro `generate` will contain the name `abc`. In the program you should have a line like

```
. generate 'generate' = ...
```

which generates a new variable abc. For more information about programming see
the volume of the *Stata Reference Manual* which is devoted to Stata programming.
In particular, see the set trace, set tracedepth, and pause commands which are
useful when debugging a program.

Exercises

1. Start by opening the file lgraph.do in the *Do–file Editor*, then add the lines to
   make it into a program, and save it as lgraph.ado in your working directory.
   Just to be sure which file is which, try

   ```
   . dir lgraph.*
   ```

   and check that you have two files, lgraph.do and lgraph.ado, in your working directory. Open lgraph.ado in the *Do-file Editor* and improve it step by
   step. After each improvement, save the file, then discard the old version of the
   program with the command discard, and run the new one.

2. Write a command which will convert a variable which contains dates as numeric
   variables of the form yyyymmdd into a new variable in Stata format. Some
   dates in the form yyyymmdd are provided in the file testdates.dta. Load this
   file and list the variable date. The easiest way to create convert a date in this
   form to Stata date format is to proceed as follows:

   (a) The year equals int(date/10000) where int() returns the integer part.
   (b) The mmdd part is equal to date − 10000 × year.
   (c) The month is equal to int(mmdd/100).
   (d) The day is equal to mmdd − 100 × month.
   (e) The function mdy(month, day, year) will convert to Stata date format.

   Try this on 19650621 and you will see that

$$
\begin{aligned}
\text{int}(19650621/10000) &= 1965 \\
19650621 - 10000 \times 1965 &= 621 \\
\text{int}(621/100) &= 6 \\
621 - 100 \times 6 &= 21
\end{aligned}
$$

   so the answer is 21 June 1965. Write a program to convert a date variable in
   the form yyyymmdd to a new variable in Stata format, and test it on the file
   testdates. Check your answer using a more general program called todate,
   contributed by Nick Cox, which is included with the files which came with the
   book.

Chapter 20

How Stata is organized

In this chapter you will learn how to use the `adopath` command; how to update Stata over the internet; how to use the Stata FAQs; how to subscribe to the Statalist; how to install commands contributed by users; and where to find the *Stata Journal*.

20.1 Paths and programs

Stata does not come as a single monolithic program which the user is unable to modify. Instead the philosophy is to allow the user as much control as possible. There is a relatively small compiled binary file which carries out the task of organizing and interpreting the rest of the software, including data input, but most of the Stata commands come as independent files to which the user can gain access. These files are called *ado files*, which stands for *automatically loaded do files*. They have the extension `.ado`, so for example, the program code for the command `table` is in the file `table.ado`. There are many hundreds of ado files which make up Stata, and they increase in number all the time. Many were written by users, and adopted by Stata Corporation after careful checking.

Each Stata ado file itself consists of Stata commands. This is why the software takes only a small amount of space on the hard disk. If each command consisted of code for a stand-alone program, the package would be many times bigger. This is also what makes it possible for users to write simple programs for their own use. A new program may consist of only a few lines of other Stata commands, whereas a stand-alone version would be much larger and writing the code would require far greater programming skills.

An important command for understanding how Stata is organized is `adopath`. This reports the paths which Stata will search when you type a command. Try

```
. adopath
```

If you are using Stata under MS Windows you should see something like this:

```
[1]   (UPDATES)    "C:\stata\ado\updates/"
[2]   (BASE)       "C:\stata\ado\base/"
[3]   (SITE)       "C:\stata\ado\site/"
[4]                "."
[5]   (PERSONAL)   "C:\ado\personal/"
[6]   (PLUS)       "C:\ado\plus/"
[7]   (OLDPLACE)   "C:\ado/"
```

If you are working on a network it might look something like this:

```
[1]   (UPDATES)    "N:\stata\ado\updates/"
[2]   (BASE)       "N:\stata\ado\base/"
[3]   (SITE)       "N:\stata\ado\site/"
[4]                "."
[5]   (PERSONAL)   "C:\ado\personal/"
[6]   (PLUS)       "C:\ado\plus/"
[7]   (OLDPLACE)   "C:\ado/"
```

where the first three entries now refer to network directories, but the rest are the same as before. Note that Stata accepts either forward slash or backward slash in pathnames. The paths are listed in the order in which they are searched. To find table.ado, Stata looks first in the UPDATES directory, to see whether the original table.ado has been updated. If it is not found there, Stata looks in the BASE directory, and so on, down the list. The entry "." stands for your working directory.

The rest of the paths are there to help you in customizing Stata. For example there may be commands shared by all users at a site, written by someone at that site, which will be stored in N:\stata\ado\site/; you may have some personal commands stored in C:\ado\personal/; or you may have downloaded some commands from the internet and installed them in C:\ado\plus/. Additional paths can be added to the search list, as in

```
. adopath + C:\courses\ado/
```

which will add C:\courses\ado/ to the search list. Similarly paths can be removed, most easily by number. For example

```
. adopath -3
```

will remove the SITE directory, and re-number the rest. It is sometimes useful to add a path to the start of the search list, to make sure that Stata looks there first. Try

```
. adopath ++ C:\courses\ado/
```

to add C:\courses\ado/ at the start of the search list. These facilities are useful because Stata can only find commands which are in directories on the adopath.

To find out in which directory a particular command, such as `table`, has been found, try

```
. which table
```

and you should see that `table.ado` is in the `BASE` directory, while

```
. which cdf
```

should show that `cdf.ado` is in the `PLUS` directory.

20.2   Updating Stata

The way Stata is organized, as a set of interlinked programs, means that it is important to update the package regularly. Fortunately, provided you have a connection to the internet, and permission to write to the Stata directory, there is a very simple way of updating Stata over the internet. The command

```
. update query
```

will report the current state of your system, and advise whether it needs updating. The advice might be to update the executable (i.e. the Stata core program), or to update the ado files, or to do both. In response to the advice you can type one of

```
. update exec
. update ado
. update all
```

A typical `update all` will take around 15 minutes on a 28.8 bps connection. For more details of how to proceed if you don't have permission to write to the Stata directory, see the Stata *Getting Started* manual.

20.3   The Stata Journal

The *Stata Journal* publishes reviewed papers together with shorter notes or comments, regular columns, book reviews and other material of interest to Stata users. For information on the journal, including instructions for authors, see

```
http://www.stata-journal.com
```

and for information on how to obtain copies see

```
http://www.stata.com/bookstore/sj.html
```

When papers in the journal describe new commands the ado files for these can be freely downloaded (see below). Prior to the *Stata Journal*, new commands were published in the *Stata Technical Bulletin* (STB), and these commands are still available from the Stata website.

20.4   User-contributed commands

User-contributed commands are supplied without any guarantee that they will work properly, but in fact are usually of a high standard. Finding a user–contributed command usually starts with a keyword search to see whether commands relevant to a particular application exist. For example, click on *Help*, then *Search*, then *Search documentation and FAQs*, and type `cumulative` in the box provided, or type

```
. search cumulative, local
```

in the command window. The search on `cumulative` first provides commands which are included in Stata, such as `cumul`. These are indicated by [U] for *User's Guide* or [R] for the *Reference Manual*. Any relevant commands which have been published by the SJ (or STB) follow. Clicking on the blue code corresponding to the command puts you in a position to install. For example, clicking on `gr37` will install the `cdf` command with its help file in the PLUS directory, first checking that it is not already installed. If it is, no action is taken, unless you ask for it to be replaced. You can also get any ancillary files (usually data files) which will go into your current working directory. To widen the search you can click on *Search net resources* or type

```
. search cumulative, net
```

in the command window. This will search a number of websites, including the SSC-IDEAS website, which contains a wealth of material. Clicking on *Search all*, or typing

```
. search cumulative, all
```

searches both the local and the net resources. An equivalent command is `findit` which presents the results in a Stata viewer window. As an example, try

```
. findit cumulative
```

What all of the these searches are looking for is Stata packages: these are collections of files – typically `.ado` and `.hlp` files – which provide a new feature in Stata. Once you have located a package, you can usually install it by clicking on the (blue) invitation to install.

In general, the commands

```
. net from      (the source URL)
. net install   (the name of the package)
. net get       (the name of the package)
```

will work for any package for which you know the URL. The `net install` installs the ado files while the `net get` loads any ancillary files. The `net` command is also useful for installing packages from your own hard disk or CD-ROM (see Chapter 0).

20.5 The Statalist

A useful resource for both beginners and experienced users of Stata is the Statalist, which is a listserver, and distributes messages to all subscribers. Although independent of Stata Corporation, the list is carefully monitored by the Corporation for problems with the current version of Stata and suggestions for the next release. It is recommended that all users join the list, which is easily done by sending the message "subscribe statalist" to

`majordomo@hsphsun2.harvard.edu`

To unsubscribe, send the message "unsubscribe statalist". To send a message to the list, send it to

`statalist@hsphsun2.harvard.edu`

but if it is a request for help, you are strongly advised to check the manual, and to read the FAQs (Frequently Asked Questions) at the Stata website before doing this. These are kept at

`http://www.stata.com/support/faq`

and are a very useful source of information.

20.6 Other sources of help

The main sources of information about Stata are the *User's Guide*, and the *Reference Manuals*[10]. These are an excellent source of information about both Stata and statistics. In addition Stata Corporation and Timberlake Consultants offer courses over the internet. See `www.stata.com` and `www.timberlake.co.uk` for details of dates and prices.

Stata has a technical support group which will sort out any problems for registered users, but before contacting them you are advised to check the FAQs and other sources of documentation. See the Stata webpage, under *User Support* and *Technical Support* for more details.

A number of useful books for learning more about statistics while using Stata have been published. See the Stata webpage under *Bookstore* for more details.

The Stata community is generous with its help. There are many sources of written material in addition to those mentioned above, but two in particular are

`http://www.ats.ucla.edu/stat/stata`

for general material, and

`http://www.iser.essex.ac.uk/teaching/stephenj/ec968/index.php`

for material on survival analysis.

Bibliography

[1] E. Christensen, J. Neuberger, J. Crowe, D.G. Altman, H. Popper, B. Portmann, D. Doniach, L. Raneck, N. Tygtrup, and R. Williams. Beneficial effect of azathioprine and prediction of prognosis in primary biliary cirrhosis. Final results of an international trial. *Gastroenterology*, 89:1084–1091, 1985.

[2] D.G. Clayton and M. Hills. *Statistical Models in Epidemiology.* Oxford University Press, Oxford, 1993.

[3] Mario Cleves, William W Gould, and Roberto Guttierez. *An Introduction to Survival Analysis using Stata, Revised Edition.* Stata Press, College Station, TX, USA, 2004.

[4] N.J. Cox. Speaking Stata: How to repeat yourself without going mad. *The Stata Journal*, 1:86–97, 2001.

[5] W.A. Guy. *Journal of the Royal Statistical Society*, 6:197–211, 1843.

[6] N. Mantel and W. Haenszel. Statistical aspects of the analysis of data from retrospective studies. *Journal of the National Cancer Institute*, 22:719–748, 1959.

[7] K. Molbak and T. Hald. An outbreak of *salmonella* typhimurium in the county of Funen during late summer. A case-control study. *Ugeskr Laeger*, 159:5372–7, 1997.

[8] J.N. Morris, J.W. Marr, and D.G. Clayton. Diet and heart: a postscript. *British Medical Journal*, 19 November(2):1307–14, 1977.

[9] K.J. Rothman. *Modern Epidemiology.* Little, Brown, and Company, Boston, 1986.

[10] StataCorp. *Stata Statistical Software: Release 8.2.* Stata Press, College Station, TX, USA, 2003.

Index